T0194659

MACAT

An Analysis of

David Riesman's

The Lonely Crowd
A Study of the Changing
American Character

Jarrod Homer

Published by Macat International Ltd
24:13 Coda Centre, 189 Munster Road, London SW6 6AW.

Distributed exclusively by Routledge
2 Park Square, Milton Park, Abingdon, Oxon OX14 4RN
711 Third Avenue, New York, NY 10017, USA

Routledge is an imprint of the Taylor & Francis Group, an informa business

www.macat.com
info@macat.com

Cataloguing in Publication Data
A catalogue record for this book is available from the British Library.
Library of Congress Cataloguing-in-Publication Data is available upon request.
Cover illustration: Alex Tomlinson

ISBN 978-1-912303-81-6 (hardback)
ISBN 978-1-912128-17-4 (paperback)
ISBN 978-1-912282-69-2 (e-book)

Notice
The information in this book is designed to orientate readers of the work under analysis,
to elucidate and contextualise its key ideas and themes, and to aid in the development
of critical thinking skills. It is not meant to be used, nor should it be used, as a
substitute for original thinking or in place of original writing or research. References and
notes are provided for informational purposes and their presence does not constitute
endorsement of the information or opinions therein. This book is presented solely for
educational purposes. It is sold on the understanding that the publisher is not engaged
to provide any scholarly advice. The publisher has made every effort to ensure that
this book is accurate and up-to-date, but makes no warranties or representations with
regard to the completeness or reliability of the information it contains. The information
and the opinions provided herein are not guaranteed or warranted to produce particular
results and may not be suitable for students of every ability. The publisher shall not be
liable for any loss, damage or disruption arising from any errors or omissions, or from
the use of this book, including, but not limited to, special, incidental, consequential or
other damages caused, or alleged to have been caused, directly or indirectly, by the
information contained within.

CONTENTS

THE MACAT LIBRARY

The Macat Library is a series of unique academic explorations of seminal works in the humanities and social sciences – books and papers that have had a significant and widely recognised impact on their disciplines. It has been created to serve as much more than just a summary of what lies between the covers of a great book. It illuminates and explores the influences on, ideas of, and impact of that book. Our goal is to offer a learning resource that encourages critical thinking and fosters a better, deeper understanding of important ideas.

Each publication is divided into three Sections: Influences, Ideas, and Impact. Each Section has four Modules. These explore every important facet of the work, and the responses to it.

This Section-Module structure makes a Macat Library book easy to use, but it has another important feature. Because each Macat book is written to the same format, it is possible (and encouraged!) to cross-reference multiple Macat books along the same lines of inquiry or research. This allows the reader to open up interesting interdisciplinary pathways.

To further aid your reading, lists of glossary terms and people mentioned are included at the end of this book (these are indicated by an asterisk [*] throughout) – as well as a list of works cited.

Macat has worked with the University of Cambridge to identify the elements of critical thinking and understand the ways in which six different skills combine to enable effective thinking.
Three allow us to fully understand a problem; three more give us the tools to solve it. Together, these six skills make up the **PACIER** model of critical thinking. They are:

ANALYSIS – understanding how an argument is built
EVALUATION – exploring the strengths and weaknesses of an argument
INTERPRETATION – understanding issues of meaning

CREATIVE THINKING – coming up with new ideas and fresh connections
PROBLEM-SOLVING – producing strong solutions
REASONING – creating strong arguments

To find out more, visit **WWW.MACAT.COM.**

CRITICAL THINKING AND *THE LONELY CROWD*

Primary critical thinking skill: INTERPRETATION
Secondary critical thinking skill: ANALYSIS

David Riesman's *The Lonely Crowd: A Study in the Changing American Character* is one of the best-known books in the history of sociology – holding a mirror up to contemporary America and showing the nation its own character as it had never seen it before.

Its success is a testament to Riesman's mastery of one key critical thinking skill: interpretation. In critical thinking, interpretation focuses on understanding the meaning of evidence, and is frequently characterized by laying down clear definitions, and clarifying ideas and categories for the reader. All these processes are on full display in *The Lonely Crowd* – which, rather than seeking to challenge accepted wisdom or generate new ideas, provides incisive interpretations and definitions of ideas and data from a variety of sources.

Above all, Riesman's book is a work of categorization – a form of interpretation that can be vital to building and communicating systematic arguments. With the aid of his two co-authors (Nathan Glazer and Reuel Denney), he defined three cultural types that formed a perfect pattern for understanding mid-century American society and the changes it was undergoing. The clarity of the book's definitions tapped directly into the zeitgeist of the 1950s, powering it to best-seller status and an audience that extended far beyond academia.

ABOUT THE AUTHOR OF THE ORIGINAL WORK

David Riesman was born in 1909 into a wealthy, secular Jewish American family. He trained as a lawyer, but at the end of World War II decided to swap his successful legal career for a life working in the emerging field of sociology. Published in 1950, his first book, *The Lonely Crowd*, looked at the evolving American social character. It was immensely popular, but its success proved a double-edged sword, raising Riesman's profile while leaving him open to heavy criticism by academics. Although the public loved the book, most seemed to misinterpret it, believing Riesman was critical of the postwar American character. He died in 2002 at the age of 92, respected but largely bypassed by today's sociologists.

ABOUT THE AUTHOR OF THE ANALYSIS

Dr Jarrod Homer took his PhD in sociology at the University of Manchester for research focusing on American culture and Jewish artists of the mid-twentieth century.

ABOUT MACAT

GREAT WORKS FOR CRITICAL THINKING

Macat is focused on making the ideas of the world's great thinkers accessible and comprehensible to everybody, everywhere, in ways that promote the development of enhanced critical thinking skills.

It works with leading academics from the world's top universities to produce new analyses that focus on the ideas and the impact of the most influential works ever written across a wide variety of academic disciplines. Each of the works that sit at the heart of its growing library is an enduring example of great thinking. But by setting them in context – and looking at the influences that shaped their authors, as well as the responses they provoked – Macat encourages readers to look at these classics and game-changers with fresh eyes. Readers learn to think, engage and challenge their ideas, rather than simply accepting them.

'Macat offers an amazing first-of-its-kind tool for interdisciplinary learning and research. Its focus on works that transformed their disciplines and its rigorous approach, drawing on the world's leading experts and educational institutions, opens up a world-class education to anyone.'

Andreas Schleicher
Director for Education and Skills, Organisation for Economic Co-operation and Development

'Macat is taking on some of the major challenges in university education ... They have drawn together a strong team of active academics who are producing teaching materials that are novel in the breadth of their approach.'

Prof Lord Broers,
former Vice-Chancellor of the University of Cambridge

'The Macat vision is exceptionally exciting. It focuses upon new modes of learning which analyse and explain seminal texts which have profoundly influenced world thinking and so social and economic development. It promotes the kind of critical thinking which is essential for any society and economy. This is the learning of the future.'

Rt Hon Charles Clarke, former UK Secretary of State for Education

'The Macat analyses provide immediate access to the critical conversation surrounding the books that have shaped their respective discipline, which will make them an invaluable resource to all of those, students and teachers, working in the field.'

Professor William Tronzo, University of California at San Diego

WAYS IN TO THE TEXT

KEY POINTS

- David Riesman (1909–2002) was an American sociologist*
  (a scholar of the history and functioning of human society)
  who grew up in the American city of Philadelphia. He
  practiced as a lawyer before turning his attention to the
  study of American society.

- *The Lonely Crowd: A Study of the Changing American
  Character* explored changes in the American social
  character and the ways in which people within a particular
  group may be similar to one another.

- While shaped as an academic study, Riesman's work
  became a best-selling book among the general population.
  It revolutionized the ways in which Americans in the
  mid-twentieth century viewed themselves and their
  contemporaries.

Who Was David Riesman?

David Riesman, the author of *The Lonely Crowd: A Study of the
Changing American Character* (1950) was born into a wealthy German
Jewish family in 1909. The family lived in Philadelphia, Pennsylvania,
where his father worked as a professor of clinical medicine. Riesman
studied biochemical science (the chemistry of biological processes) at
Harvard University, then law at Harvard Law School (while also

managing to study government, economics, and history). He received his law degree in 1934 and was admitted to the bar in Massachusetts.

After World War II,* however, Riesman decided on a career in academia. He was offered a lectureship at the University of Chicago and worked at Yale University, where he began work on *The Lonely Crowd*.

Little from the author's early life and career suggested that he would one day become one of the most successful sociologists of his generation. That said, there are autobiographical elements evident in *The Lonely Crowd* that explain this shift. In the book Riesman discusses changes in the social character of Americans. For him, US society before the war, oriented towards industrial production, was largely inner-directed*—that is, behavior was governed by internal values and beliefs that had been instilled early in life. Postwar, Riesman saw a change. He argues that Americans were becoming other-directed:* behavior was increasingly governed by the behavior and dictates of external groups, such as their peers or the media. The struggle between these two types, and the more general struggle to locate autonomy (independence of action), authenticity, and individual identity, is the interest that drives the study. It could be that in his own life Riesman moved from an inner-directed identity (his ideals and goals driven by familial trends) to an other-directed identity, as an aspiring sociologist. Perhaps Riesman was trying to articulate a personal but nevertheless common social experience.

What Does *The Lonely Crowd* Say?

The Lonely Crowd is a sociological analysis of post-World War II social identity in the United States. Riesman argues that historically the American social character can be best described as inner-directed. Inner-directed individuals have an inner understanding about what they should and should not do, of morals and ambitions. These values, deeply embedded when the individuals are young, help create and maintain a dynamic social atmosphere, which drives industrial expansion. In the postwar era, however, a variety of changes occurred. The structure of American society shifted from being a society based on production to a society based on consumption, prompting a change in social character. With it, an other-directed* personality type emerged. Instead of having an inner sense of who they are and what they stand for, they navigate a confusing social landscape (full of options and opportunities) by responding to external influences.

The Lonely Crowd was a best seller in the 1950s. The book has gone on to sell more than 1.5 million copies. Its popularity is a testament to the way in which Riesman tapped into the zeitgeist (the spirit of the time) of the mid-twentieth-century United States. The public could relate to the types of social characters he was describing. As a result, *The Lonely Crowd* was used to articulate the private and public preoccupations of a curious and anxious American citizenship in the 1950s—a time of dramatic change when Americans were eager to find answers to new social problems. A generation of newly suburban Americans were furnishing their homes, changing their leisure and religious practices, adjusting to new work patterns, and joining the expanding middle class. *The Lonely Crowd* was used as a tool—a form of self-help guide—to help this generation understand the shifts happening around them.

The popular success of *The Lonely Crowd* had not been anticipated. Riesman had written his book for an academic audience. But the response of the academic world was far more critical than that of the public. Some of Riesman's insights were accepted as astute. These include the links he draws between the self and the state, between individuality* and conformity* and between personal taste and mass media* (any means of communication that can reach a large number of people, such as television, radio or newspapers).

While these insights may be applied to any culture at any time, Riesman's methods drew an onslaught of criticism. He was also criticized for the ambition of his approach. In writing *The Lonely Crowd* Riesman had drawn on a huge variety of sources and had explored issues from across a number of disciplines. Many critics argued that his understanding in many of these disciplines was weak. This criticism has intensified over the years, as Riesman's broad, ambitious approach has fallen out of favor in the field of sociology. Nevertheless, despite its varied flaws, the impact that Riesman's study had in popular culture and a number of scholarly disciplines in the mid-century secures the status of *The Lonely Crowd* as a deeply influential text.

Why Does *The Lonely Crowd* Matter?

In many ways, Riesman's insights have proved prescient—that is, somewhat prophetic. The author was among the first social commentators to identify a movement away from a society and economy based on production to one based on consumption. He looked at the links between the socioeconomic structure of society and how social character is shaped and defined. In a nation whose culture and global brand would go on to be characterized by drive-ins, malls, and coffee shops, this is an important and universal aspect of Riesman's study.

Riesman was also among the first to conceptualize the notion of the peer group,* an influence group made up of a person's associates, friends, and contemporaries. A peer groups mediates influences from mass-media sources. Riesman was keen to link this notion to youth culture in particular. He highlighted the ways in which the peer group acted as an agent of socialization. It could appraise social trends and could evaluate and sometimes impact upon another key social influence: mass media.

Riesman saw the relationship between the media, the peer group, and the individual as complex. While this relationship could certainly encourage conformity* (obedience to shared norms), the peer group could—Riesman suggested—be a means by which individuals could gain political power and influence. The peer group could bring individuals together as a democratizing force: a pressure group. Similarly, mass media could be harnessed as a communicating tool. Today's world of social media networking and technological advances has enabled both consumption and community. As a result, Riesman's ideas about the emergence of both conformity and peer-bonded pressure groups would appear to be among his more pertinent and lasting insights.

Riesman's work continues to matter for its insights. It also remains relevant as a text that helped to popularize the field of sociology. The success of *The Lonely Crowd* helped establish sociology as a distinct area of study. In addition to this, the book remains of immense value to historians of the United States in the 1950s. It offers a valuable insight into the desires, anxieties, and fears that troubled the American psyche during the mid-twentieth century.

SECTION 1
INFLUENCES

MODULE 1
THE AUTHOR AND THE
HISTORICAL CONTEXT

KEY POINTS

- David Riesman (1909–2002) was an American sociologist*
  (a scholar of the history and functioning of human society)
  who grew up in the American city of Philadelphia. He
  practiced as a lawyer before turning his attention to the
  study of American society.

- *The Lonely Crowd: A Study of the Changing American
  Character* explored changes in the American social
  character and the ways in which people within a particular
  group may be similar to one another.

- While shaped as an academic study, Riesman's work
  became a best-selling book among the general population.
  It revolutionized the ways in which Americans in the
  mid-twentieth century viewed themselves and their
  contemporaries.

Why Read This Text?

David Riesman's *The Lonely Crowd: A Study of the Changing American
Character* was a phenomenally popular book. Aimed at an academic
audience, it received a lukewarm reception when it was first published
in 1950. But its fortunes changed when the paperback edition came
out in 1953: the book proved an unexpected hit with the public,
selling more than half a million copies by 1955. The ideas in *The Lonely
Crowd* were discussed across the United States, and had a substantial
influence. In particular, the book created an informal understanding of
a person's place within the social hierarchy. The sheer popularity of
Riesman's study secures it as a key text in popular sociology. In his

> **❝** [*The Lonely Crowd*] was based on our experiences of living in America—the people we have met, the jobs we have held, the books we have read, the movies we have seen, and the landscape. **❞**
>
> David Riesman, *Faces in the Crowd*

description of social character*—the ways in which people within a particular group may be similar to one another—and in his description of the contemporary American cultural experience* he undoubtedly captured the mood of his time.

While there were other influential American sociological books written in the 1950s, *The Lonely Crowd* should be considered an especially influential and groundbreaking text. It appeared to predict the way in which the social character of most Americans formed in the mid-twentieth century. And its foresight extends beyond the sociological makeup of 1950s mainstream culture in the United States. In addition to its scientific study of society and social behavior, *The Lonely Crowd* also touches on areas of interest to economists and psychologists. Riesman foresaw the socioeconomic shift from production to consumption in the US. He predicted the increasingly blurred lines between the private and the public and between work and leisure. And he stressed the importance of other-directedness* in contemporary America—the ways in which an individual's identity is forged in reference to external influences, such as his or her peers or the media. This has relevance in today's world of social media and online consumer culture.

Author's Life
Riesman was born in 1909, into a wealthy, refined, and assimilated German Jewish family. He was raised and remained secular, and enjoyed the benefits of a good education and a privileged early life.

Graduating from Harvard in 1931 as a biochemist (a scholar of the chemistry of biological processes), he returned three years later to earn a degree in law. As a well-connected and promising young lawyer, he worked as a clerk for the Supreme Court Justice Louis D. Brandeis* and at a Boston law firm, before going on to teach law at the University of Boston.

In 1941 he took a temporary leave of absence from Boston to accept a fellowship at the Columbia Law School. Before he could return to Boston, however, the US entered World War II* and the school there was shut down. During this time Riesman worked as a deputy district attorney in New York and for the Sperry Gyroscope* Company. A gyroscope is a spinning wheel, held in a frame that allows it to turn in any direction while maintaining its stability. One of *The Lonely Crowd's* most memorable metaphorical motifs is Riesman's comparison of the inner-directed* character—an individual who is guided by internal autonomy and inner values and beliefs—with the self-stabilizing gyroscope.

Riesman began his switch to sociology after the war, taking up a position as assistant professor at the University of Chicago in 1946. After a few years in Chicago he took leave to pursue a project at Yale University, investigating the mid-century American character. To help him, Riesman recruited the American sociologist Nathan Glazer* and the American poet and academic Reuel Denney.* The result of the work was *The Lonely Crowd*.

Author's Background

Riesman was a man with numerous connections. He met, associated with, or worked alongside a great number of thinkers, academics, intellectuals, and critics who would provide inspiration and insight for his own ideas. During his time in New York during the war, for example, Riesman associated with leading social scientists such as Margaret Mead,* Ruth Benedict,* Robert S. Lynd,* and Helen

Merrell Lynd.* The depth and diversity of Riesman's intellectual curiosity is evident in *The Lonely Crowd*.

One of his key inspirations was the German social psychologist* Erich Fromm.* Fromm gained considerable popularity by writing extensively on psychoanalysis,* a therapeutic and theoretical approach to the unconscious workings of the human mind. From 1939 Riesman attended Fromm's clinic in Manhattan to undergo private psychoanalysis. This personal connection, along with Fromm's writing, greatly influenced Riesman in *The Lonely Crowd*. The American historian Daniel Horowitz* describes this process: "Riesman both absorbed and transformed what he learned from Fromm and other émigré intellectuals. Turning away from legal studies, he combined European critical theory with American traditions of social criticism based on empirical social science research."[1]

Although Riesman is the chief author, researcher, and driving critical force behind *The Lonely Crowd*, his two coauthors, Nathan Glazer and Reuel Denney, were crucial in its writing and research and to the study's overall success. A lawyer by training and education, Riesman was an enthusiastic but nevertheless novice sociologist. He did not have the requisite qualifications in the field to go it alone. Glazer and Denney offered valuable subject expertise and experience. Glazer, a sociologist, helped Riesman to define the character types discussed in the book, and helped to explore their historical provenance. Reuel Denney, a poet, academic, and social analyst, embellished Riesman's understanding of the experience of contemporary American youth and added to Riesman's discussion of sports and popular music.

NOTES

1   Daniel Horowitz, "David Riesman: From Law to Social Criticism," *Buffalo Law Review* 58 (2010): 1005.

MODULE 2
ACADEMIC CONTEXT

KEY POINTS

- Riesman uses the ideas of the French political thinker of the nineteenth century, Alexis de Tocqueville,* to place his exploration of contemporary American social character within a broader historical context.
- *The Lonely Crowd* draws on diverse sources to explore its subject.
- The book drew its many and varied sources into a study offering a new way of understanding private and public social interaction.

The Work in its Context

In *The Lonely Crowd: A Study of the Changing American Character* (1950), David Riesman uses a wide range of sources, from a range of disciplines. He explores personal testimony and popular culture.

The French political thinker Alexis de Tocqueville's nineteenth-century analysis of America informs the work. So do many mass-media texts and population surveys of the twentieth century. Riesman draws on the therapeutic and theoretical approach of psychoanalysis,* founded by the Austrian neurologist Sigmund Freud,* and on the socioeconomic insights of the American economist Thorstein Veblen.* As one reviewer points out, "The arguments of *The Lonely Crowd* were profound and original, but they took place within the context of earlier debates over individualism, scarcity and abundance."[1]

> ❝ It is my impression that the middle-class American of today is decisively different from those Americans of Tocqueville's writings who nevertheless strike us as contemporary, and much of this book will be devoted to discussing those differences. ❞
>
> David Riesman, *The Lonely Crowd: A Study of the Changing American Character*

Other sources include the following:
- a large body of literature relating to sociology and psychology
- population surveys that explore voting patterns, leisure pursuits, and work habits
- interviews with US citizens
- interpretations of popular culture in the form of films, literature, plays, radio programs, and poetry
- investigations into mass-media journalism.

The book is informal and ambitious and does not necessarily interact with prior sociological studies. One way of looking at the text, therefore, is to see Riesman's intellectual environment as entirely personal and of his own making. *The Lonely Crowd* can be read as a culmination of the author's interest in a number of academic disciplines and his curiosity about the American way of life.

Overview of the Field

In *The Lonely Crowd*, Riesman did not set out to *disprove* any existing thesis or understanding; his text acts, rather, as a meeting point for various ideas and disciplines. He draws on many of the ideas of Alexis de Tocqueville, the French political thinker who visited the United States in 1831. Tocqueville's book *Democracy in America* analyzed the nation's cultural and political life in the nineteenth century. Relying heavily on Tocqueville for an understanding of that period, he uses his idea that an increasing tendency toward conformity creates a lack of individualism in American social life. Riesman links this to insights of contemporary sociologists and psychologists.

The German social psychologist* Erich Fromm* (a scholar of the ways in which society, social behavior, and the human mind act on each other) identified the "marketer" type, for example: characters who measure their self-identity through their own ability to "sell" themselves as a commodity. The American sociologist C. Wright Mills* identified the "fixer"* type: people who use their own intelligence, initiative, and guile to forge new avenues of enterprise. For Riesman, both are figures created by a dominant mass culture (an internally consistent culture that serves to promote typical ideas and values).

No one thinker dominates Riesman's work, however. Tocqueville, Mills, Fromm, and many others form part of the rich background to the ideas set out in *The Lonely Crowd*.

The book was very different from much of the discussion regarding mass culture in the mid-century. This was especially true of the studies that followed the work of the German sociologist Theodor W. Adorno,* one of the most important thinkers in Germany in the postwar period. Adorno argues that mass culture is an inescapable and harmful monolith. Riesman recognizes that mass culture does have the potential to engender conformity* (an acceptance of shared norms), but unlike Adorno, he sees mass culture—or at least mass media—as offering opportunities for increased individualism.* Mass media enables the individual to move past his or her peer group* and become aware of a wider world and more choice. Abundance, consumer culture, and advertising offer flexibility and autonomy (the capacity to act on behalf of oneself). In this way mass media may counteract the conformity and asceticism (austere self-denial) of the workplace and of conservative sociopolitical ideology.

Academic Influences

It is problematic to try to position *The Lonely Crowd* within an intellectual battleground that existed prior to the text's publication. But there were other works published around the same time that looked at the relationship between individual and collective identity. *White Collar: The American Middle Classes* (1951), by the American sociologist C. Wright Mills investigated the relationship between the middle class and a postwar "power elite."[2] *The Organization Man* (1956) by the journalist William Whyte* emphasized the role of the workplace in forming social character.[3] And Riesman's ideas were informed by the work of the German social psychologist Erich Fromm. Fromm's books *Escape from Freedom* (1941) and *Man for Himself: An Inquiry into the Psychology of Ethics* (1947), helped form Riesman's understanding of both the relationship between individual autonomy and personal freedom, and the sociological rigors of the modern environment.[4] In *Man for Himself* Fromm discusses the "market orientation" of the middle classes. He believes this creates a media-influenced individual who finds "conviction of identity not in reference to himself and his powers but in the opinion of others about him. His prestige, status … success are a substitute for the genuine feeling of identity."[5]

In *The Lonely Crowd*, Riesman echoes this with his discussion of the movement from inner-direction* to other-direction.* The inner-directed individual bases her or his behavior on internal beliefs and values. The other-directed individual bases his or her behavior on standards set by others such as peer groups or consumer culture.

The Lonely Crowd was an important text in the developing discipline of social studies, which emerged in America during the mid-century and gained widespread popularity beyond academia. In this context we can see Riesman at the beginning of a movement that saw sociology enjoy a period of heightened relevance within American popular culture.

NOTES

1   Joseph Featherstone, "John Dewey and David Riesman: From the Lost Individual to the Lonely Crowd," *John Dewey: Political Theory and Social Practice* 2 (1992): 59.

2   C. Wright Mills, *White Collar: The American Middle Classes* (New York: Oxford University Press, 1951).

3   William Hollingsworth Whyte, *The Organization Man* (New York: Simon & Schuster, 1956).

4   Erich Fromm, *Escape from Freedom* (New York, Rhinehart, 1941); *Man for Himself: An Inquiry Into the Psychology of Ethics* (London: Routledge, 2013). *Escape from Freedom* was published as *The Fear of Freedom* in the UK.

5   Fromm, *Man for Himself*, 73.

MODULE 3
THE PROBLEM

KEY POINTS

- *The Lonely Crowd* uses a multilayered understanding of the term "social character":* the traits and features that define a social group. It examines the idea of American society within a historical and a contemporary cultural context.

- It was one of the first studies in the burgeoning discipline of social science.

- While the book sparked much debate, Riesman was not concerned with proving or disproving existing ideas or theories.

Core Question

David Riesman's seminal book *The Lonely Crowd: A Study of the Changing American Character* (1950) was one of the first studies in the discipline of sociology,* helping to popularize and establish the field as a distinct area of study. As such, the text stands alone, rather than developing from existing ideas or responding to an active critical debate. As Riesman says, "My collaborators and I base ourselves on this broad platform of agreement, and do not plan to discuss in what way these writers differ from each other and we from them."[1]

The Lonely Crowd's ideas had their genesis in various disciplines. Riesman's study was a complex interweaving of material from numerous fields, including psychology (the study of the human mind and behavior), anthropology (the study of human beings, especially the cultural and social behavior that defines us as such), and history. These were either applied to ideas about American social character

> **❝** Riesman was a pioneer in the development of sociology as literature. He had an immense, omnivorous curiosity, a capacious temperament that made him open to a broad spectrum of cultural experiences. He drew on his observations of a wide range of material— children's books, movies, novels, interviews, and social science data. **❞**
>
> Daniel Horowitz,* *David Riesman: From Law to Social Criticism*

and history, or fused with other sources such as population studies, voting data, and analysis of the mass media.* This allowed Riesman to discuss how contemporary postwar behavioral patterns and social experience could be understood in a broader historical perspective.

The mid-century was a time of great social change. Americans were moving to the suburbs. They were encouraged to furnish their homes and consume readily. Increasingly, an emphasis was placed on the concept of US identity, fuelled by an ever-growing and ever more influential mass media and mass cultural* experience. In *The Lonely Crowd*, Riesman was answering a cultural need. At a time when Americans wanted to understand themselves, Riesman's study asked some important questions about what it meant to be an American.

The Participants
Riesman acknowledged his indebtedness to a large number of scholars and thinkers. For example, in his discussion of "social character" he says that the concept will be familiar to readers acquainted with the writings of the German social psychologist Erich Fromm,* the American anthropologist Abram Kardiner,* or the American cultural anthropologist Margaret Mead.*[2]

These were not the only scholars from whom Riesman drew inspiration. It would be difficult to list all of the influences that were brought to bear on Riesman and *The Lonely Crowd.* Among others, he also touched on the ideas and theories of the nineteenth-century French political thinker Alexis de Tocqueville* and the German-born American developmental psychologist Erik H. Erikson.* Riesman never seeks to disprove the work of other scholars. He uses their writing as theoretical evidence to support his own ideas about social character. *The Lonely Crowd* does not participate in debate, then; it exists as a coming-together of various sources, theories, and ideas to inform a wholly original thesis and discussion.

The Lonely Crowd is a richly textured study that benefits from Riesman and his coauthors' connections and associations with some of the leading thinkers in sociology and psychology. It also benefits from their vast reading of historical literature and their sophisticated (and exhaustively researched) understanding of the contemporary cultural climate.

The Contemporary Debate

The authors of *The Lonely Crowd* wanted to understand and categorize American social character in the mid-twentieth century. Taking a historical perspective, they looked at how social character had changed over time. The study can be seen as part of a contemporary debate about American social and national identity. Riesman's work influenced the books of popular sociology that were published after *The Lonely Crowd*. Many of the trends and areas of interest addressed by his text were built upon and analyzed by subsequent authors in the 1950s. The United States in the post-World War II* era became increasingly anxious about the social character of the nation.

The Lonely Crowd might be compared to the work of the American sociologist C. Wright Mills.* Mills's book *The Power Elite* (1956) discusses the notion of sociopolitical power. In it, he argues that power is maintained by the few.[3] Riesman suggests a more pluralistic understanding. He argues that the other-directed*—those whose behavior is directed by the values and beliefs of others, such as their peers or the media—are rewarded for their correct comprehension of social structures and practices. *The Lonely Crowd* could also be considered alongside a book by the American journalist Vance Packard,* *The Status Seekers* (1959).[4] Both books discuss how 1950s American citizens were acutely aware of their social position in the context of those around them. They explore how things like consumption practices act as markers of class and social identity.

NOTES

1   David Riesman, Nathan Glazer, and Reuel Denney, *The Lonely Crowd: A Study of the Changing American Character* (New Haven: Yale University Press, 2001), 4.

2   Riesman, *The Lonely Crowd*, 4.

3   C. Wright Mills, *The Power Elite* (New York: Oxford University Press, 1956).

4   Vance Packard, *The Status Seekers: An Exploration of Class Behavior in America* (Harmondsworth: Penguin, 1961). The book was first published in 1959 in Philadelphia by D. McKay Co.

MODULE 4
THE AUTHOR'S CONTRIBUTION

KEY POINTS

- Riesman's chief concern in *The Lonely Crowd* is to outline, explain, and contextualize his two key concepts of social character:* the other-directed* and the inner-directed.*

- *The Lonely Crowd* created a model that Americans could use to assess their own social standing and general social character. It also inspired a decade of popular social scientific inquiries.

- The book was both a cultural and a critical revelation. It offered a new conception of American social character.

Author's Aims

It is important to note that in *The Lonely Crowd: A Study of the Changing American Character* (1950) David Riesman did not have an ideological agenda. He was viewing American culture through the inquisitive eye of a social anthropologist. He writes, "This is a book about ... the way in which one kind of social character, which dominated America in the nineteenth century, is gradually being replaced by a social character of quite a different sort. Why this happened; how it happened; what are its consequences in some major areas of life: this is the subject of this book."[1]

Riesman offered a historical perspective on the relationship between different social characters. More specifically, he located the American postwar character in the context of this history in order to explain how conformity*—compliance with contemporary social norms—functioned in the postwar United States. Riesman was interested in why this mode of conformity and social character differed

> ❝ David Riesman called for the individual to acquire 'autonomy.' Rather than becoming a revolutionary who was opposed to society, individuals had to resist the compulsion to accept the opinions of others. ❞
>
> Mark Jancovich, *Rational Fears*

from previous historical types of social character. He asked what the consequences of this might be for everyday lived experience, and how it might affect the overall cultural climate in America.

How successful was Riesman in doing this? The answer to that is complicated by the popularity of his book with the general public. The book was intended for an academic community. Its wider success brought sociology into mainstream cultural consciousness. While this encouraged the development of sociology and popular science, it meant that in discussion Riesman's thesis was simplified to make his arguments more accessible—which led to the book being misinterpreted. It was assumed that the book mourned the movement toward other-directedness* and the loss of an inner-directed* social character (one that was incorrectly assumed to be free to act independently). In this way *The Lonely Crowd* and Riesman became associated with a critique of American culture that warned against conformity.

While this misinterpretation was quite outside of Riesman's control, the book's popularity alters our perception of how well he achieved his original aims.

Approach

In 1953, when *The Lonely Crowd* was published as a paperback, it quickly became popular. Widely read, Riesman's study was called upon to answer a number of questions of urgent interest to many Americans: Who are we? Why do we act the way that we do? Where are we heading as a society?

These are questions that often emerge in the midst of widespread social upheaval and a large-scale demographic shift. The way in which the text struck a chord with Americans in the mid-twentieth century is a testament to the timely and astute nature of the author's central inquiry. Riesman set out to understand how and why social character is formed and what dominant social characteristics might be. He wanted to understand the ways in which social character impacted upon the psychological, political, cultural, and economic experiences of contemporary mid-century Americans. He also wanted to know how these impacted on the consumption practices, work patterns, and leisure pursuits of contemporary Americans.

The Lonely Crowd became almost immediately indispensable within contemporary American social discourse. That fact serves to highlight the originality of the approach taken by Riesman and his collaborators. It also underlines their perceptiveness in producing a study that met a real and urgent public need for such a work.

Contribution in Context

Riesman's early career was as a lawyer. Before the publication of *The Lonely Crowd*, he was already a prolific writer, his work dealing primarily with issues related to law. Given the difference in disciplinary focus it is difficult to see similarities between his earlier work and *The Lonely Crowd*. But even at this time Riesman was showing an interest in the public sphere and class distinctions—as they related to legal issues such as defamation, insult, and political abuse. This suggests an early interest in the notion of how cultural groupings impact on public behavior.

More pertinent to Riesman's later work is his "The Meaning of Opinion," (1948), written with the American sociologist Nathan Glazer,* one of the coauthors of *The Lonely Crowd*.[2] "The Meaning of Opinion" was produced at Yale University. In it, Riesman and Glazer sought to analyze public opinion polls. They looked at the connections between different answers in polls and different groups of people. They also explored what a respondent's answer really "means." As the authors put it, "finally, we must find ways to relate these responses to character structure, if we are to attempt to predict political or other behaviors."[3]

The authors wanted to explore how social character impacts on private and individual behavior. They were also exploring the relationship between an individual's external and internal character. In addition, they were studying how social and individual identity can be seen as dependent upon a variety of political and cultural influences. This early paper shows that Riesman's collaboration with Glazer helped to form the intellectual interests of *The Lonely Crowd*—social categorization (the "kind" of social being one might be thought to be), opinion, character formation, and social behavior.

NOTES

1   David Riesman, Nathan Glazer, and Reuel Denney, *The Lonely Crowd: A Study of the Changing American Character* (New Haven: Yale University Press, 2001), 3

2   David Riesman and Nathan Glazer, "The Meaning of Opinion," *The Public Opinion Quarterly* vol. 12 no.4 (1948–49): 633–48.

3   Riesman and Glazer, "The Meaning of Opinion," 633.

SECTION 2
IDEAS

MODULE 5
MAIN IDEAS

KEY POINTS

- *The Lonely Crowd* discusses the ways in which an individual's freedom to act is restrained by both internal and external authority.

- Riesman is interested in the impact of society upon the day-to-day lives of the individuals who live in that society.

- The popular success of *The Lonely Crowd* is due, in part, to Riesman's easy-to-read writing style.

Key Themes

David Riesman's seminal work *The Lonely Crowd: A Study of the Changing American Character* (1950) is structured around one main theme: the impact society has on the day-to-day life of an individual. Society encourages social and psychological conformity* to the norms of that society. This, in turn, discourages an individual's autonomy, or freedom to act. For Riesman, there are three different social characters:

- Tradition-directed:* A tradition-directed individual is shaped by deeply embedded social traditions. These are social expectations and notions of social propriety that are universally accepted and unquestionable.

- Inner-directed:* An inner-directed individual behaves in accordance with her or his own, internal set of values and beliefs. These beliefs are cultivated at an early age, giving the individual a seemingly innate knowledge of what a person should or should not do. Riesman says, "The source of direction for the individual is 'inner' in the sense that it is implanted early in life by the elders and directed toward generalized but nonetheless inescapably destined goals."[1]

> ❝ This is a book about social character and about the
> differences in social character between men of different
> regions, eras, and groups. It considers the ways in which
> different social character types, once they are formed at
> the knee of society, are then deployed in the work, play,
> politics, and child-rearing activities of society. ❞
>
> David Riesman, *The Lonely Crowd: A Study of the Changing American
> Character*

- Other-directed:* An other-directed individual behaves in
  accordance with the cultural consensus. These people are shaped
  by the cultural climate formed by their peer group,* the media,
  and the social norms of the day. Their character is dependent
  upon a wealth of external forces. They find personal orientation
  within a confusing social landscape by taking cues from both
  peers and mass-media influences. Riesman elaborates: "What is
  common to all the other-directed people is that their
  contemporaries are the source of direction for the individual."[2]

Riesman suggests that social character shifts over time. It is affected
both by the way in which authority is expressed within a given culture
and by social and economic factors.

Exploring the Ideas

For Riesman, social character is formed by the socioeconomic and
political imperatives of the day. It is shaped by who holds authority
within a given society, and how that authority is expressed. In turn,
social character leads to different forms of social behavior and different
social atmospheres.

Tradition-directed social character is associated with societies that
place a strong emphasis on familial and communal bonds: such as a

premodern environment and developing nations. This type of social character helps to maintain deeply engrained social hierarchies, organized along clear boundaries of caste, age, or clan. A tradition-directed environment discourages individual autonomy in favor of maintaining long-held attitudes and solutions to create a stable but static society.

Riesman links inner-direction with historical periods in which individual endeavor is paramount. For example, Riesman saw the period of European history known as the Renaissance*—an era in which practitioners of the arts and architecture turned toward ancient Greek and Roman forms to reinvigorate European culture—as a time when artistic endeavor marked the transition between the medieval world and the modern world. Encouraging inner-directedness inspires self-reliance and a rigidity of character that enables the individual to navigate through a changing social climate.

The other-directed type is at home in a society that de-emphasizes political, social, and economic enterprise among its citizens and prioritizes social cooperation. In this environment industrial innovation has largely been achieved and the frontiers of society are more fixed. One such society was the post-World War II* United States, with its rapidly growing middle class.*

Riesman argues that other-directedness has shaped every aspect of society in the United States in the years following World War II. Individual behavior is now primarily shaped by the horizontal authority exerted by the mass media and peer groups, as opposed to the top-down authority of family and lineage. Riesman explores the breadth of social experience in the United States of the period and contrasts this with older forms of social organization. In doing this he shows how social character has developed over time.

Language and Expression

While the subject of Riesman's inquiry created widespread interest, it was Riesman's writing style and his use of language that made the book accessible and appealing to a mass audience. As the American sociologist Todd Gitlin* explains in the foreword to the 2001 edition of the book, "*The Lonely Crowd* was lucidly written, with a knack for puckish phrases … It was decidedly unpretentious, unforbidding in tone, omnicurious, with a feeling for recognizable types … it had the sound of an agreeable human voice, by turns chatty and approachably awkward, graceful and warm, nuanced and colloquial, sober and avuncular, but frequently casual and good-humored. Unlike most academic treatises, it did not get bogged down in definitional character."[3]

The first edition of *The Lonely Crowd*, published in 1950, was met with a welcoming if lukewarm critical reception. However, the 1953 paperback edition sold in the hundreds of thousands. This popularity with such a large audience was unexpected. And as general interest in *The Lonely Crowd* grew so, too, did the text's popularity among academics. More and more sociological studies of the middle classes and their consumption practices and social experiences were published in the years following the publication of Riesman's text.

NOTES

1   David Riesman, Nathan Glazer, and Reuel Denney, *The Lonely Crowd: A Study of the Changing American Character* (New Haven: Yale University Press, 2001), 15.

2   Riesman, *The Lonely Crowd*, 21.

3   Todd Gitlin, foreword to *The Lonely Crowd: A Study of the Changing American Character*, by David Riesman, Nathan Glazer and Reuel Denney (New Haven: Yale University Press, 2001), XIII.

MODULE 6
SECONDARY IDEAS

KEY POINTS

- *The Lonely Crowd* addresses the relationship between population patterns and social character.

- The book contained many interesting secondary ideas that have not received much attention. Chief amongst these are Riesman's ideas about how individuals can have autonomy within a mass culture,* when cultural norms are broadcast through the media and other cultural outlets. "Mass culture" is an internally consistent culture that serves to promote widely accepted ideas, ideals, and values.

- The development of social networking, online communities, and a culture of online reviewing have proven Riesman's insights about the relationship between peer groups* and the mass media* to have been somewhat prophetic.

Other Ideas

The Lonely Crowd: A Study of the Changing American Character by David Riesman is a complex study, containing many ideas, notions, and theories. Some are explored thoroughly, some are merely suggested and abandoned. Riesman's work is fairly informal in this respect, as the author tries to feel his way through to an understanding of the United States' social character. Chief among Riesman's secondary ideas is the observation that social character is directly related to population patterns. This was actually intended to be part of the central hypothesis of *The Lonely Crowd* and is important because it underpinned much of Riesman's reasoning about how and why the United States shifted from predominantly inner-directed* behavior to predominantly other-directed* behavior.

> 66 The idea that men are created free and equal is both
> true and misleading: men are created different; they lose
> their social freedom and their individual autonomy in
> seeking to become like each other. 99
>
> David Riesman, *The Lonely Crowd: A Study of the Changing American
> Character*

Another secondary idea worth mentioning is the idea of individual
autonomy within an other-directed* culture. Riesman attempts to
suggest ways for individuals to harness the mass media and politics and
explores ways in which the individual might obtain authority from the
dominant social order and alter dominant patterns of social behavior.

Riesman discusses how the political landscape is colored by the
dominant social order of other-directedness. Like Riesman's ideas
regarding autonomy, this discussion places the issue of social character
formation within an urgent, real-world setting. As a result, these
chapters can perhaps be seen as more instructive than the previous
sections of the book. Writing in 1950, Riesman accurately predicts the
social experience of Americans in the decade that followed. *The Lonely
Crowd* foretells, for example, the relationship between politics and
consumership (the state of being a consumer). The book also discusses
veto-groups: pressure groups that Riesman argues will be powerful
enough to block policies that go against their interests. In doing this,
he suggests that more groups will gain influence in decision-making.

Exploring the Ideas

The three distinct social types identified in *The Lonely Crowd* are
linked to specific historical moments and cultural climates and are
associated with behavioral patterns and social interactions. They are
also based on a population model:* that is, they are connected to an
understanding of birth-and death-rate trends in the population.

Riesman argues that the different stages of social character—tradition-directed,* inner-directed, and other-directed—are determined by different population rates. At times when the population is stable and the social order is static, with little technological advancement, a tradition-directed social type tends to dominate. This type helps to maintain the existing social order for generations. In contrast, during times of large population growth, technological innovation, and industrial expansion, inner-directed social types flourish. This climate fosters the virtues of individual endeavor, ambition, and social enterprise. The inner-directed social type seeks self-betterment by following an inner authority built in childhood and occurs in societies geared toward production.

In the post-World War II* period a culture of consumption and commercialism replaced a culture based on production. For Riesman, this cultural trend dovetailed with reduced population growth. With frontiersmanship* (skills of self-sufficiency) no longer an imperative in this society, the social character became less dynamic.

The population model is a questionable methodological approach, as Riesman was aware in 1949 when his study was in the final stage of preparation for publication.[1] The problem for his hypothesis is that the post-World War II period witnessed a drastic increase in population, a baby boom* that counters Riesman's connection between other-directed* social character and reduced population growth. The faults with his model do not wholly undermine the study—Riesman's observations regarding historical and post-World War II social types remain largely true—but the link to movements in population was quickly dismissed.

Overlooked

Attempts have been made by numerous commentators to counter the popular interpretation of Riesman's work. Rather than viewing Riesman as a straightforward and somewhat negative critic of consumerism, conservatism, and mass culture, these reviewers see *The Lonely Crowd* as offering useful advice on how to maintain autonomy within a mass-cultural environment. They have tried to redirect understanding of *The Lonely Crowd*, so that what Riesman intended to say becomes better understood. They have also tried to highlight the utility of his work.

The sociologist Neil McLaughlin,* for example, champions Riesman's analysis of children as trainee consumers in an other-directed social climate. He argues that the study "is not simply a well-written academic bestseller, but it is also a serious attempt to theoretically integrate Freudian* social theory into American empirical sociology" ("empirical" here meaning, roughly, sociology that draws on observation and not wholly on theory). He also discusses, more generally, Riesman's important attempt to synthesize social understanding with critical and theoretical reasoning.[2]

More recently in the *New Yorker*, the American writer Gideon Lewis-Kraus* illustrates the finer points of Riesman's ideas relating to the mass media and autonomy.[3] He wrote the article "Yelp and the Wisdom of The Lonely Crowd"

in response to a *New York Times* article by the American writer Lee Siegel,* which had used a misreading of *The Lonely Crowd* as a straightforward condemnation of conformity* and unthinking crowd mentality.

NOTES

1   Todd Gitlin, foreword to *The Lonely Crowd: A Study of the Changing American Character*, by David Riesman, Nathan Glazer and Reuel Denney (New Haven: Yale University Press, 2001), xv.

2   Neil McLaughlin, "Critical Theory Meets America: Riesman, Fromm, and The Lonely Crowd," *The American Sociologist* vol. 32, no. 1 (2001): 16.

3   Gideon Lewis-Kraus, "Yelp and the Wisdom of *The Lonely Crowd*," The *New Yorker*, May 7, 2013, accessed June 15, 2013, www.newyorker.com/online/blogs/elements/2013/05/the-wisdom-of-the-lonely-crowds.html.

MODULE 7
ACHIEVEMENT

KEY POINTS

- Riesman set out to map a timeline of American social character*; despite the shortcomings of his study, he succeeded.
- The cultural insights of *The Lonely Crowd* helped to popularize the book.
- *The Lonely Crowd*'s "life" as a popular self-help guide obscured many of Riesman's more prescient ideas.

Assessing the Argument

The popularity of David Riesman's book *The Lonely Crowd: A Study of the Changing American Character* came at a cost. Many of the author's more complex ideas were missed by the popular interpretation of the book, according to which *The Lonely Crowd* was ideologically against mass culture* and social conformity. Riesman was seen as a critic of the conformist and consumerist direction that American culture was taking in the mid-twentieth century. It was believed that his work mourned the loss of an individual and autonomous American character. Although Riesman tried to correct this misinterpretation in subsequent editions, *The Lonely Crowd* acquired a life of its own. It became a popular vehicle for critiquing and understanding the mid-century American social experience.

Over time, academics responding to *The Lonely Crowd* have found themselves in a predicament. On the one hand much of the text's methodological, historical, and theoretical approach and understanding has been proven incomplete or incorrect. As a result, the book is not of much interest to serious sociologists or other academics. Its audience

> ❝ [The] book made 'inner-direction' and 'other-direction' household terms, canapés for cocktail party chat. It was read by student radicals in the making, who over interpreted its embrace of the search for autonomy as a roundhouse assault on conformity ... Its title phrase even cropped up in a Bob Dylan* song of 1967. ❞
>
> Herbert J. Gans,* "Best-sellers by American Sociologists: An Exploratory Study"

here is based on enthusiastic reappraisals that forgive its shortcomings. These scholars focus on the general way in which the book marries sociology and theories of social consciousness.

On the other hand, the book's usefulness as a popular cultural text relies heavily upon its misappropriation, making *The Lonely Crowd* less a work of sociology and more a cultural artifact. It gives historians and humanities scholars a better understanding of the concerns of the mid-century American mindset; in this way the dialogue between the text's original intent and its cultural misappropriation may prove useful for understanding aspects of American culture in the 1950s, such as masculine anxiety.

Achievement in Context

As a part of the popular cultural consciousness and having captured the zeitgeist (the spirit of the time), *The Lonely Crowd* became a popular reference point. In the mid-twentieth century the United States underwent a dramatic demographic shift (a change in the makeup of its population). The country also moved away from a production-based society to one based around the notion and practice of consumption.

Riesman's book led to informal quasi-intellectual discussions at social gatherings around the country. Americans decided to which group they and their associates belonged. *The Lonely Crowd* gave voice to a generation of newly suburban Americans as they worried their way through the 1950s. This was the decade in which a generation of consumers furnished their homes, changed their leisure and religious practices, adjusted to new patterns of work, and joined the expanding middle class.

The Lonely Crowd appeared to predict the cultural trends and social patterns of the era. It became, for some, an indispensable guide to understanding the social interactions and the cultural terrain of the 1950s. The book discusses other-directed* characters seeking ways in which to measure their own individual character. It is, therefore, somewhat ironic that *The Lonely Crowd* became a tool for helping individual Americans understand their own place, and the place of their peers, within the social character of the 1950s.

Limitations

There were a number of best-selling sociological studies in the post-World War II* era. These include *The Power Elite*, by the American sociologist C. Wright Mills;* *The Organization Man* by the journalist William Whyte* and *The Status Seekers*, by the journalist Vance Packard.*[1] Riesman's study stands alongside these as a work that helped to establish sociology as a discipline. *The Lonely Crowd* went some way toward establishing a useful approach within sociology: the dialogue between the analysis of social organization and experience and theoretical reasoning.

Certain aspects of *The Lonely Crowd* might have prompted more serious debate if the text had not become so popular with a general audience. Riesman's study was misinterpreted as a straightforward critique of mass culture rather than a serious inquiry about social-psychological character. This obscured some of its arguments. It was damaged further by criticism of its incorrect or incomplete methodological, historical, and theoretical approach and understanding.

The insights that should have gained further attention include Riesman's identification of the movement away from a socioeconomic structure based on industrial production to one based upon consumption. Riesman was also among the first to conceptualize the notion of a peer group—a group made up of one's associates, friends, and contemporaries that mediated influences from mass-media* sources.

Riesman's methodology has fallen out of fashion in sociology today. His approach blends analysis of cultural forms and messages with empirical sources (sources verifiable by observation), and marries a theoretical idea of "consciousness" with a social and psychological investigation that perhaps exceeds his theory's capacity to provide explanations. In addition, Riesman takes a broad and ambitious interest in the everyday lived experience.

Today, this methodology has been superseded by an emphasis on proving narrow hypotheses using a scientific and data-focused approach, moving sociology away from examining wider cultural and sociopolitical questions regarding social character, identity, and experience. The popular sociology of thinkers such as Riesman, Daniel Bell,* C. Wright Mills and Vance Packard is now confined to the past. Gone are the early days of the field when largely untrained sociologists attempted interpretive analyses of American culture, often accompanied by an ideological or political bias.

NOTES

1   C. Wright Mills, *The Power Elite* (New York: Oxford University Press, 1956); William Hollingsworth Whyte, *The Organization Man* (New York: Simon & Schuster, 1956); Vance Packard, *The Status Seekers: An Exploration of Class Behavior in America* (Harmondsworth: Penguin, 1961).

MODULE 8
PLACE IN THE AUTHOR'S WORK

KEY POINTS

- David Riesman's primary approach throughout his life's work was to offer a sympathetic but critical appraisal of American social character.*

- *The Lonely Crowd* was Riesman's first book. It eclipsed all of his other work.

- The book made Riesman a household name in the 1950s, putting him on the front cover of *Time* magazine.

Positioning

The Lonely Crowd: A Study of the Changing American Character (1950) is, without doubt, the most important text of David Riesman's career. The book encapsulates Riesman's critical preoccupations: the desire to explain everyday experience and to understand it within a historical context. But *The Lonely Crowd* was not Riesman's only work; he published widely and enthusiastically in a variety of disciplines. Before and during World War II* Riesman drew on his early career as a lawyer, publishing articles and papers on various aspects of law, such as civil liberties. He published a number of influential articles on the psychology of defamation and slander.

After *The Lonely Crowd* was published Riesman became a commentator on American higher education. In 1968 he produced—with Christopher Jencks*—an important work about the rise of professional scholars: *The Academic Revolution*.[1] Generally speaking, other of his works share the same interests as *The Lonely Crowd*. *Individualism Reconsidered and Other Essays* (1966) and *Abundance for What? and Other Essays* (1964) focus on American culture and explore

45

> **" *The Lonely Crowd*, marked the beginning of an age of bestselling readership for a handful of academics attempting to gauge the temper of a burgeoning consumer society. "**
>
> Paul Buhle,* "Obituary: David Riesman," The *Guardian*

social behavior and character.[2] Riesman's body of work is united by his desire to analyze and critically appraise American society. In *Individualism Reconsidered* Riesman echoes his idea about the other-directedness* of American culture from the mid-twentieth century onward. He states, "What is feared as failure in American society is, above all, aloneness. And aloneness is terrifying because it means there is no one, no group, no approved cause, to submit to."[3]

Integration
The Lonely Crowd was one of Riesman's earlier books; its popularity tends to overshadow his other work. It was a speculative and ambitious project and brought Riesman to the attention of millions.

The book fits into Riesman's overall output in a fairly narrow but important way. He returned to the analysis and conclusions of *The Lonely Crowd* in a book published with Nathan Glazer* in 1952— *Faces in the Crowd: Individual Studies in Character and Politics*.[4] This text sought to expand and secure the central ideas of *The Lonely Crowd*, which was itself republished in 1961 and 1969. Riesman wrote a preface to both reprints in which he explained the genesis of the original text and addressed some of the concerns that had been voiced about the first version of *The Lonely Crowd*.

Riesman maintained his profile as a noted sociologist. He also continued to publish widely and gained a reputation in other areas before his death in 2002, at the age of 92. As well as the massive cultural impact made by *The Lonely Crowd*, Riesman achieved lasting success as a commentator on American higher education. He was a respected liberal academic with research interests in politics, education, American cultural identity, and the plight of the individual, among other topics.

Significance

The Lonely Crowd's longevity depends on two things:

- Its informal use in popular culture—the popular reading of Riesman's text sees it as a critical appraisal of unthinking conformity* and the negative effect that unfettered consumerism* (the culture of purchasing) has on selfhood and autonomy.
- The book's role as a marker of the myth that depicts the 1950s as an era of consumerism and conformity.

As the American intellectual historian Wilfred M. McClay* describes: "The book's greatest and most enduring strengths are cautionary ones. It warns us against the peculiar forms of bondage to which our era is especially prone. And in doing so, it draws us into a deeper consideration of what freedom might be, both now and in the future. For that reason, it may well endure for another 50 years. Or even longer."[5]

Riesman's study asked some important questions at a time when American life was changing and when a greater emphasis was being placed upon American identity. In this sense the timeliness of *The Lonely Crowd* was so particular and apt that its success would be hard to replicate. It is for this reason that the book dominates Riesman's career and academic output.

The Lonely Crowd is estimated to have sold more than 1.5 million copies to date.[6] This is despite the problems with the original text, many of which have been admitted by Riesman—if not necessarily corrected. Riesman's understanding of American society and the relationship between society and character has proved to have lasting resonance. *The Lonely Crowd* popularized the field of sociology and the work of scrutinizing American society.

NOTES

1   Christopher Jencks and David *Riesman, The Academic Revolution* (Garden City, NY: Doubleday, 1968).

2   David Riesman, *Individualism Reconsidered and Other Essays* (Glencoe, Illinois: Free Press, 1954); *Abundance for What? and Other Essays* (Garden City, NY: Doubleday, 1964).

3   David Riesman, *Individualism Reconsidered and Other Essays* (Glencoe: Free Press, 1954).

4   David Riesman and Nathan Glazer, *Faces in the Crowd: Individual Studies in Character and Politics* (New Haven: Yale University Press, 1952).

5   Wilfred M. McClay, "Fifty Years of *The Lonely Crowd*," *The Wilson Quarterly* vol. 22 no. 3 (Summer 1998): 34–42, accessed 15 October, 2015, http://archive.wilsonquarterly.com/sites/default/files/articles/WQ_VOL22_SU_1998_Article_02.pdf.

6   Todd Gitlin, foreword to *The Lonely Crowd: A Study of the Changing American Character* by David Riesman, Nathan Glazer and Reuel Denney (New Haven: Yale University Press, 2001), XII.

SECTION 3
IMPACT

MODULE 9
THE FIRST RESPONSES

KEY POINTS

- Criticism of *The Lonely Crowd* came from all quarters. Critics questioned the author's understanding of the numerous disciplines, theories, and scholarly debates from which his study borrows.

- Riesman accepted criticism of his work. He responded to it by reemphasizing the most important aspects of his study, reiterating his original intentions in writing the book, and conceding to his critics on various issues.

- Despite the criticism leveled at *The Lonely Crowd*, its popularity within the mid-century culture is a testament to the accuracy of Riesman's appraisal of American society.

Criticism

Reviews of David Riesman's book *The Lonely Crowd: A Study of the Changing American Character* appeared in journals of psychology, psychiatry, politics, literature, and sociology, as well as in popular academic journals like *American Quarterly* and *Social Problems*. The originality of Riesman's insights and his far-reaching approach impressed reviewers. But appreciation of Riesman's thesis was tempered with criticism of his approach, his reasoning, and the depth of his understanding in a variety of fields.

The Lonely Crowd was of interest to people working in a wide range of disciplines. The breadth of the work suggests that Riesman's study would lack depth in some areas. To the casual reader the spread of interests provide a series of intriguing insights into modern life. But to the expert in any given field the investigation was spread too thinly, leaving important gaps.

> **❝** *The Lonely Crowd* was greeted in professional journals with often quite astringent criticism, and it made its way only slowly to a wider, nonprofessional audience. **❞**
>
> David Riesman, *The Lonely Crowd* (Twenty Years After—A Second Preface)

In 1961 the American political sociologist Seymour Martin Lipset* edited, with Leo Lowenthal,* *Culture and Social Character: The Work of David Riesman Reviewed*, a collection of the criticism of Riesman's work.[1] It offers an excellent summary view of the ways in which *The Lonely Crowd* was criticized throughout the 1950s. Lipset attacks Riesman's overly simplistic historical understanding, which ignores aspects of older American social characters. The American sociologist Talcott Parsons* criticizes the attention Riesman gives to the psychological aspect of social character. He argues that this serves to de-emphasize cultural patterns and norms. The American sociologists Robert Gutman* and Dennis H. Wrong,* meanwhile, take aim at Riesman's understanding of the psychoanalytical* approach, which, they argue, is incomplete; for them, what Riesman calls social characters are merely opposing value systems. And the American sociologist Norman Birnbaum* rejects Riesman's suggestion that, within politics, there is a trend towards pluralism— shared, rather than elitist power. Asking whether Riesman gives "sufficient weight to postwar interaction of internal and foreign politics," Birnbaum suggests that there are times when Riesman "adopts an astonishingly simple view."[2]

Responses

While Lipset and Lowenthal's *Culture and Social Character: The Work of David Riesman Reviewed* was a collection of criticism of *The Lonely Crowd*, Riesman also contributes to it, offering revisions and explanations of his conclusions and methodology.[3] In addition, in a lengthy preface to the 1961 edition of *The Lonely Crowd*, Riesman takes the opportunity to address some of the issues around the text. He also provides some background to the study and its intellectual context. Conceding to the criticism on the population model, Riesman explains, "in 1949 we felt that we should not try at the last minute to take account of this issue, but simply to present the population hypothesis as an interesting but unproved idea."[4]

Riesman also concedes some of the criticism about his conceptualization of veto-groups (pressure groups that Riesman argues will be powerful enough to block policies that go against their interests). In *The Lonely Crowd* Riesman asserts that this form of political power acted as a successful counterpoint to the power elite—individuals with an abundance of political influence. In responding to the criticism, he maintains that his general assertions remain relevant, but that there had been some changes in power relations since *The Lonely Crowd* was published. Riesman also discusses his text's treatment of mass media.* He argues that the mass media terrain had shifted since 1950. His hopes about the democratizing effects of television and the news media had not been realized. But he adds that *The Lonely Crowd* might be used as an instructive text for those who might want to reform the media: "We think that the book might help those who want to reform the media to go about it in a more intelligent way. For one thing, we encouraged people to discriminate among the media: and today we would point out that the movies are less cowardly than they were when we wrote."[5]

Conflict and Consensus

In the editions of *The Lonely* Crowd that followed its original publication in 1950 we are left with an almost unchanged text. In his prefaces to the editions of 1961 and 1969 Riesman nevertheless recognizes the book's shortcomings. The 1961 preface offers a clear assessment of where his study might be revised, while in the 1969 preface Riesman offers new insights, discussion, and perspective on the book.

Drastically revising *The Lonely Crowd* would certainly have improved and developed the study's conclusion. However, by referencing his critics and responding to their charges while leaving his original analysis largely intact, Riesman encourages debate about the subject of his study. This approach allows *The Lonely Crowd* to act as a point of reference for those interested in the development of mass media and the understanding of social character. It also allows the book to remain a useful cultural artifact in its own right, marking a time when anxieties over social character were rife. Although there were serious shortcomings with the methodology of *The Lonely Crowd*, and with some of its conclusions, this did little to reduce its importance as a cultural text. In addition, Riesman makes sure in his prefaces that the criticism the book received was used to redirect the reader to the most important aspects of his study.

NOTES

1   Seymour Martin Lipset and Leo Lowenthal, eds., *Culture and Social Character: The Work of David Riesman Reviewed* (New York: Free Press, 1961).

2   Norman Birnbaum, *Toward a Critical Sociology* (Oxford University Press, 1973), 220. All criticisms are to be found in Lipset and Lowenthal, *Culture and Social Character*.

3   Lipset and Lowenthal, *Culture and Social Character*.

4   David Riesman, Nathan Glazer, and Reuel Denney, *The Lonely Crowd: A Study of the Changing American Character* (New Haven: Yale University Press, 2001), LIII.

5   Riesman, *The Lonely Crowd*, LXV.

MODULE 10
THE EVOLVING DEBATE

KEY POINTS

- Riesman was the first and the best known of the "Big Notion" thinkers who gained popularity in mid-twentieth-century American culture.

- The "Big Notion" thinkers examined the American mindset and cultural character of the mid-twentieth century.

- *The Lonely Crowd* lives on as a cultural artifact that gave voice and guidance to an anxious middle America as it adapted to new social patterns in the 1950s.

Uses and Problems

Within the academic field of sociology* David Riesman's book *The Lonely Crowd: A Study of the Changing American Character* quickly lost its import. As the sociologist Waltraud Kassarjian surmises, "In spite of the wide interest in systematic theories of social processes and social behavior, research employing or testing Riesman's social character types is extremely scarce … Riesman himself in subsequent writings gives no additional research data [while various critics] attempted to study only one or the other aspect of inner-other-directedness as it may be related to some outside factor."[1]

With its population model disproved and its methodology and understandings called into serious question, Riesman's study fell out of favor. It no longer possessed the critical edge, the evidential platform (a body of reliable evidence on which to base its claims), or the methodological accuracy necessary to be a reliable sociological study. Riesman had failed to offer discussion of race or gender in the book. In addition he had used a social-psychological* model instead

> **❝** Sociologists seldom cite David Riesman in peer
> reviewed journal articles today. And he has often been
> dismissed as a popularizer and as part of the discipline's
> pre-professional history. **❞**
>
> Neil McLaughlin, "Critical Theory Meets America: Riesman, Fromm, and
> *The Lonely Crowd*"

of a social-structural* model ("social psychological" here being a
model in which individual thoughts, feelings, and behaviors are
directly influenced by the external or imagined presence of a
person's peers and other members of society; "social-structural"
being an understanding of society as formed of structures such as
class groups and social institutions that dictate and react to the
behaviors and actions of individuals).

Riesman focused on how an individual's thoughts, feelings, and
behavior are influenced by other people. But sociology as a discipline
was more interested in the effect on behavior of social structures (like
class) and of social institutions. As a result, Riesman's study does not fit
the demands of current scholarly trends.

Riesman's brand of sociological inquiry bears little resemblance to
the intellectual world of contemporary sociology—part of the reason
why he does not have any disciples today, and why there are no
scholars that use his findings or his methodology within the field of
sociology.

Schools of Thought

Riesman, alongside other sociologists like C. Wright Mills,* Vance
Packard,* William H. Whyte,* Philip Rieff,* Daniel Bell* and
Herbert Marcuse,* published a number of hugely successful studies of
popular social criticism. These books probed the American mindset of

the mid-twentieth century. These studies were ambitious, often ambiguous, and sometimes speculative.

Discussing the trend for these works, an article in the *New York Times* in 2002 suggested, "In midcentury America, sociologists for a while rivaled … psychiatrists in their seeming ability to explain everything about everything … The 50s and 60s were the great age of the Big Notion: the pseudoscientific book that explained us to ourselves and told us where we were headed, which in most cases was nowhere good."[2]

The Lonely Crowd was the most successful of these studies. But the popularity of this style of book suggests a nation eager to accept "expert" help in analyzing itself. This was a period of sociopolitical upheaval brought about by the nuclear standoff between the United States and the Soviet Union* known as the Cold War* and post-World War II* domestic prosperity. This new prosperity led to consumer abundance, suburban sprawl, and changes in work and leisure patterns and behaviors. To help make sense of the new social environment Americans turned to these texts. They gave Americans insights into their relationships with the community, the workplace, their peers, and the ideological and political atmosphere. That helped them to orient themselves within the new social framework.

In Current Scholarship

Riesman's work offers insights into contemporary movements in American politics, social organization, and new media. But the author and his study hardly feature within current sociological thought and methodological approach.

The sociologist Neil McLaughlin,* however, argues that "Riesman's relatively neglected theoretical approach has much to offer a sociology concerned with retaining its links to public debate and empirical evidence."[3] McLaughlin argues that "the genius of *The Lonely Crowd* was to raise larger critical questions about modern culture while allowing the analysis to be moderated by … 'sociological realism' … [Riesman was] a utopian* thinker who seriously engaged the speculations and theoretical insights of great European thinkers."[4] Discussing the usefulness of Riesman's theoretically informed sociology, McLaughlin speculates that Riesman has fallen out of favor in the discipline since the 1970s. A radical sociology emerged during that period—and Riesman did not fit into it.

McLaughlin is not alone in mourning the loss of Riesman's brand of sociology. This was an approach that asked big questions about a broad social experience and that had a populist inquiry supported by an ambitious, wide-ranging philosophy and analysis. The American sociologist Todd Gitlin* also "longs for approximately ambitious, germane studies of today's mentalities—books with the reach and approachability of *The Lonely Crowd* … One wonders … how the current boom (and attendant anxieties) are playing out in the consciousness of Americans."[5]

The Jamaican-born American sociologist Orlando Patterson* notes that it is Riesman's qualities, "that independence, that confidence in ideas, that is most lacking in the academy now."[6] Having argued that sociology today is concerned with testing hypotheses via a scientific, data-focused approach, Patterson writes, "Mainstream sociology eschews any exploration of human values, meanings and beliefs because ambiguities and judgment are rarely welcomed in the discipline now."

NOTES

1   Waltraud Marggraff Kassarjian, "A Study of Riesman's Theory of Social Character," *Sociometry* vol. 25 no. 3 (1962): 213–30, accessed December 15, 2015, doi:10.2307/2786125.

2   Charles McGrath, "The Lives They Lived; Big Thinkster," The *New York Times*, December 29, 2002, accessed June 15, 2013, www.nytimes.com/2002/12/29/magazine/the-lives-they-lived-big-thinkster.html.

3   Neil McLaughlin, "Critical Theory Meets America: Riesman, Fromm, and *The Lonely Crowd*," *The American Sociologist* vol. 32, no. 1 (2001): 5.

4   McLaughlin, "Critical Theory Meets America," 20.

5   Todd Gitlin, foreword to *The Lonely Crowd: A Study of the Changing American Character*, by David Riesman, Nathan Glazer and Reuel Denney (New Haven: Yale University Press, 2001), xviii.

6   Orlando Patterson, "The Last Sociologist," *New York Times*, May 19, 2002, accessed June 15, 2013: www.nytimes.com/2002/05/19/opinion/the-last-sociologist.html.

MODULE 11
IMPACT AND INFLUENCE TODAY

KEY POINTS

- *The Lonely Crowd* anticipated key changes in society, such as the emergence of youth culture, mass media,* and peer groups.*

- The work asks readers to consider the factors that are the key influences on individual choice and personal autonomy.

- Despite having been out of favor in academic circles for some time, the *The Lonely Crowd* is experiencing a small renaissance as people explore how Riesman's ideas might help us make sense of Internet culture.

Position

When David Riesman, the author of *The Lonely Crowd: A Study of the Changing American Character*, died in March 2002, the liberal British newspaper the *Guardian* published an obituary. This ended by saying that Riesman "was a model liberal for the post-war age of the socially critical intellectual."[1] This highlights the fact that Riesman's reputation as a sociologist and his work's relevance to the field is confined to the immediate post-World War II* period.

Writing just after Riesman's death, the Jamaican-born American sociologist Orlando Patterson,* professor of sociology at Harvard University, mourns the loss of his one-time mentor. Riesman worked as a lecturer at Harvard for 30 years, and Patterson offers a sad estimation of the author's current place in the field of sociology: "David Riesman died discarded and forgotten by his discipline. [This] dishonoring [of Riesman] and the tradition of sociology for which he stood, is not a reflection of their insignificance. It is merely a sign of the

> ❝ [The] open reader returns to *The Lonely Crowd*
> feeling many aftershocks of recognition. By the turn
> of the twenty-first century, the alert observer is made
> aware every day that the shift Riesman discerned in the
> educated upper-middle classes of metropolitan centers
> has swept the country. ❞
>
> Todd Gitlin,* Foreword to *The Lonely Crowd: A Study of the Changing
> American Character*

rise in professional sociology of a style of scholarship that mimics the methodology and language of the natural sciences—in spite of their inappropriateness for the understanding of most areas of the social world."[2]

The sociologist Robert T. Sandin* also argues that sociology as a discipline has moved toward a culture "controlled by a method of thinking and research that is centered on the verification of hypotheses of limited scope by reference to statistical data generated by measurements presumed to be valid."[3] The sociology practiced by Riesman—a sociology that examined the lived experience, the beliefs, value systems, and social identity* of American culture and searched far and wide for its evidence and analysis—is no longer popular or accepted within the discipline.

Interaction

The American political scientist Alan Wolfe* has called for a reincorporation of the type of sociology seen in *The Lonely Crowd*. Wolfe sees *The Lonely Crowd* as an integral component of the "golden age of social criticism." This golden age was characterized in the mid-1950s by sociological texts that attempted a quasi-scientific interpretation of American culture and that achieved popular success.

Wolfe legitimizes the work and approach of Riesman and of the trend of social criticism to which he belonged and contributed. Aware of its flaws, Wolfe does not precisely advocate a return to its methods, recommending instead that current sociology learns from the mistakes of yesteryear and tries to incorporate its critical individualism.* As Wolfe puts it, sociology should build on "a faith in social science, a belief in democracy, and a strong sense of political commitment."[4]

Wolfe wrote this call to arms for modern sociology in 1996. But although there hasn't been a resurgence in interest in the "golden age" approach, others in the field have echoed Wolfe's support of Riesman. The sociologists Orlando Patterson and Neil McLaughlin,* for example, both champion a return to Riesman's ambitious and confident approach to sociology, admiring the way he brings together psychological and theoretical insights with perceptive social insight.

The Continuing Debate

The Lonely Crowd has had two "lives." On the one hand Riesman's study was a popular cultural text that captured the imagination of a curious and anxious American public. On the other it was a flawed piece of scholarship—albeit one that encouraged important debates across academic disciplines and helped to establish the discipline of sociology.

Today, *The Lonely Crowd* is important for what it reveals about the United States in the 1950s. In this it stands alongside other cultural sources, such as the film *12 Angry Men* (1957), directed by Sidney Lumet,* or the novel *The Man in the Grey Flannel Suit* (1955) by the American writer Sloan Wilson.*

The Lonely Crowd also lives on in popular culture. Its terms were adopted into the mainstream vocabulary in the mid-1950s as a series of buzzwords and into popular consciousness in the form of vague understandings of social types. A recent exchange between two American writers, Lee Siegel* of the *New York Times* and Gideon

Lewis-Kraus* of the *New Yorker*, exemplifies the enduring relevance of *The Lonely Crowd*. Siegel raises the idea of the "Yelpification" of culture—the way an Internet consensus and the digital community guide consumer desire and taste. He uses Riesman's study as an example of when "conformist" was "a slur on someone's character. Now the idea is that if you are not following the crowd of five-star dispensers, you're a tasteless, undiscriminating shlub."[5] In response Lewis-Kraus sets out to show that *The Lonely Crowd* was more complex than this. Riesman's analysis *championed* the notion of mass consumer choice offered by Yelp* and other opinion-sharing sites. Riesman saw mass consumer choice as a way of increasing the opportunity for consumer power and autonomy within social character.[6]

NOTES

1   Paul Buhle, "Obituaries: David Riesman," *Guardian*, May 13, 2002, accessed June 15, 2013, www.guardian.co.uk/news/2002/may/13/ guardianobituaries.socialsciences.

2   Orlando Patterson, "The Last Sociologist," The *New York Times*, May 19, 2002, accessed June 15, 2013, www.nytimes.com/2002/05/19/opinion/ the-last-sociologist.html.

3   Robert T. Sandin, "Reflections on David Riesman (1909–2002)," *The Pietisten* vol. XVII, no. 1, accessed June 15, 2013, www.pietisten.org/ summer02/riesman.html.

4   Alan Wolfe, *Marginalized in the Middle* (Chicago: University of Chicago Press, 1996), 3.

5   Lee Siegel, "Go the Same Way, or Go the Wrong Way," The *New York Times*, May 3, 2013, accessed June 15, 2013, www.nytimes.com/2013/05/05/ fashion/seeking-out-peer-pressure.html?pagewanted=all&_r=0.

6   Gideon Lewis-Kraus, "Yelp and the Wisdom of *The Lonely Crowd*," The *New Yorker*, May 7, 2013, accessed June 15, 2013, www.newyorker.com/online/ blogs/elements/2013/05/the-wisdom-of-the-lonely-crowds.html.

MODULE 12
WHERE NEXT?

KEY POINTS

- *The Lonely Crowd* remains important as a cultural artifact of the mid-twentieth century in American culture. It offers crucial insights into the understanding of the American psyche, and, as a result, will continue to be significant in the future.
- The book was an important milestone in the creation of sociology* as an academic discipline.
- Riesman's ideas about mass media,* peer groups,* and youth culture continue to be relevant in a cultural landscape dominated by social media and Internet communities.

Potential

David Riesman's book *The Lonely Crowd: A Study of the Changing American Character* is a snapshot of a particular moment in American history. Published in 1950, the book captures a time when the uncertainty of the immediate post–World War II* period combined with an anxious look towards the future. Many of Riesman's predictions about the future of the United States—ideas about the direction of American culture, consumerism,* leisure practices, work patterns, political power, and the role of the mass media—did not come to pass. As a result *The Lonely Crowd* may be read today as a hopeful work, suggesting an alternative vision of the United States' future: one that was possible before a political elite and the mass media began to tighten their grip on daily life. This, however, does not offer much of a basis for future scholarship.

> **❝In an age where many educated Americans are preoccupied with the nature of their own identities and values, many non-professional readers have come to *The Lonely Crowd* for clues as to what they are like and how they might live. ❞**
>
> David Riesman, 1961 Preface to *The Lonely Crowd: A Study of the Changing American Character*

In strictly academic terms the study also remains limited. This is because the methodological basis of Riesman's work, along with his historical framework and theoretical understanding, have all proved incorrect or incomplete. The American historian William Palmer,* for example, shows that in his historical understanding Riesman "depended too heavily on conceptions of the Middle Ages, early modern Europe, and Jacksonian* America that can no longer be accepted without serious qualification."[1] ("Jacksonian" here refers to the seventh US president Andrew Jackson, in office between 1829 and 1837.) As a result, *The Lonely Crowd* lacks credibility as a text of continued relevance to the social sciences.

Future Directions

Martin Halliwell,* a professor of American studies, uses Riesman's ideas about social character to critique a novel of the same era, J. D. Salinger's* *Catcher in the Rye* (1951). Halliwell argues that the teenage protagonist of Salinger's book, Holden Caulfield,* engages with the same problem that Riesman addressed: how to achieve autonomy in the 1950s United States. Caulfield's quest for autonomy in the adult world requires him to subscribe to a "phony" middle-class identity.[2] Halliwell is using *The Lonely Crowd* as a valid sociological study for testing the mid-twentieth century social character. He is also using Riesman's book as a cultural artifact in its own right. Halliwell places *The Lonely Crowd* alongside not only Salinger's novel but also alongside a best seller of 1955, Sloan Wilson's* novel *The Man in the Grey Flannel Suit.*

The Lonely Crowd marked a particular moment in American cultural history. It also provided an idiosyncratic interpretation of its own time, functioning as a cultural self-help book. In this regard, Riesman's study helps current scholars tap into the preoccupations of the 1950s and the sociopolitical experience of Americans in the mid-century. *The Lonely Crowd* is particularly useful as a cultural source: as well as revealing the concerns of its era, the book also helped to influence the culture of the 1950s. It is, therefore, a crucial text for understanding other mid-century discussions of American social identity.*

Summary
By the middle of the 1950s *The Lonely Crowd* had sold over half a million copies. It would go on to sell in excess of 1.5 million by last count. While popularity and large sales alone do not qualify a study as a seminal text, they can be indicative of how far a book strikes a chord at a specific moment in cultural history. That is the case with Riesman's *The Lonely Crowd.*

The success of *The Lonely Crowd* is a testament to how well the study articulated and influenced the zeitgeist* of the period. It was a time when anxieties over social identity were rife and the quest for guidance from accessible sociological texts was a popular cultural activity. Riesman's study was the most successful social text in an era that saw a particular interest in popular social criticism; these studies helped Americans to locate themselves within the changing social structure of the mid-century.

As a piece of scholarship *The Lonely Crowd* was ambitious and speculative—but it no longer stimulates serious debate, nor can it be considered a part of modern sociology. Its methods and scope are too far-reaching. Even in its contemporary scholarly environment the text invited criticism and revision. But it is because of its ambition and breadth that *The Lonely Crowd* should be considered seminal.

Riesman started a discussion about the contemporary United States that engaged both the academic world and the public. Despite its misinterpretation amongst the public and its much-criticized position in scholarship, *The Lonely Crowd* articulated ideas about the social climate of the mid-century indispensable to our understanding of that era.

NOTES

1   William Palmer, "David Riesman, Alexis de Tocqueville and History: A Look at *The Lonely Crowd* after Forty Years," *Colby Quarterly* vol. 26, no. 1 (1990): 27.

2   Martin Halliwell, *American Culture in the 1950s* (Edinburgh: Edinburgh University Press, 2007).

GLOSSARY

GLOSSARY OF TERMS

Baby boom: an era of increased birth rates often occurring at times of social upheaval (as after wartime) or economic affluence (as in the post–World War II United States).

Cold War: a period (around 1947–91) of economic, military, political, ideological, and cultural antagonism between the superpowers, the United States and the Soviet Union. The hostility stopped short of the full military action that characterizes a "hot" war.

Collectivism: a term describing something that denies individualism, while empowering the individual as part of a larger sociological group.

Conformity: behaving in accordance with the prevailing norms, social standards, attitudes, beliefs, and values of a given culture.

Consumerism: increased material consumption that drives and supports the economy. This is usually encouraged by advertising and social norms.

Cultural experience: the ways in which social identities and characters are created and influenced by mass and popular culture.

Developing nations: a term often used to describe states in Africa, Asia, and the Americas that have not achieved a level of prosperity or industrialization comparable to Western nations.

Fixer: someone who uses their own intelligence, initiative, and guile to forge new avenues of enterprise.

Frontiersmanship: characteristic of a person who lives on a frontier—often sparsely or only recently populated—who must rely on his or her individual endeavor to survive.

Individualism: the pursuit of individual interests rather than those that might be of benefit to wider society.

Inner-directed: a set of beliefs in which an individual bases their behavior, thoughts, and morals on their own set of internal values and beliefs.

Jacksonian: descriptive of the US politician Andrew Jackson, president of the United States in 1829–37, and of his brand of politics, which championed the cause of the common man.

Marketers: characters that measures their self-identity though their own ability to "sell" themselves as a commodity.

Mass culture: a homogenized culture that serves to engender typical ideas and values and that promotes mainstream ideals. The music, art, film, et cetera that constitutes mass culture is often considered to lack individuality.

Mass media: a means of communication that can reach and have influence on a large number of people: cinema, television, radio, or newspapers. Mass media tends to propagate the interests and ideals of mass culture. As such it tends towards a mainstream perspective that reflects the dominant social ideology.

Middle class: a social class that occupies the place on a class structure between upper and lower/working classes. Usually characterized by middle-income professional and nuclear family units.

Other-directed: individuals who look to an outward social character and cultural consensus to define themselves. This cultural consensus is created by factors like peer groups, the media, and consumer culture.

Peer group: a group of people with whom an individual associates and who form that individual's sphere of influence—the peer group is usually united by one or more factors such as age, social status, background, social identity, beliefs, and so on.

Population model: a model that allows for conclusions and interpretations to be draw from the dynamics and characteristics of population statistics.

Psychoanalysis: a means of understanding the unconscious psychological processes of a human being through an investigation intended to determine their influence on conscious thought and actions.

Renaissance: a cultural movement that saw a revival in art, literature, and learning in Europe between the fourteenth and seventeenth centuries. The Renaissance is generally understood to mark the transition from the medieval world to the modern world.

Social character: the aggregate or overriding traits and features of a social group.

Social identity: the identity that an individual assumes or adopts as a member of wider society.

Social norms: behavior dictated and guided by a prevailing and predominant set of beliefs, cultural and consumption practices, ideologies, and so on. These are enforced by general consensus and by wider factors like the media and advertising.

Social psychology: investigation into the ways in which individual thoughts, feelings, and behaviors are directly influenced by the external or imagined presence of a person's peers and other members of society.

Social structural: a society in which social structures (primarily class) and social institutions simultaneously dictate and react to the behaviors and actions of individuals.

Sociology: the scientific study of society and social behavior.

Soviet Union: the Union of Soviet Socialist Republics (USSR) was a Eurasian empire that arose from the Russian Revolution in 1917 and consisted of Russia and 14 satellite states in Eastern Europe, the Baltic and Black Seas, and Central Asia, existing from 1922 to 1991.

Sperry Gyroscope: a major US company that makes electronic products, now called the Sperry Corporation.

Taylorism: a scientific approach to workplace and management structure that maximizes industrial and commercial efficiency. It is named after Frederick Taylor (1856–1915).

Tradition-directed: the condition of being primarily influenced by generational rituals, routines, and kinships. This character type dominates in primitive or highly structured (class or caste-based) social organizations.

World War II (1939–1945): a global conflict fought between the Axis Powers (Germany, Italy, and Japan) and the victorious Allied Powers (United Kingdom and its colonies, the former Soviet Union, and the United States).

Utopia: a state, community, or environment that is highly desirable and perfect.

Yelp: a social review site on the Internet where users review local businesses.

Zeitgeist: the prevailing cultural and social mood of a certain period of time.

PEOPLE MENTIONED IN THE TEXT

Theodor W. Adorno (1903–69) was a German intellectual who arrived in the United States in the mid-1930s, having fled the threat of Nazism in Europe. Upon arriving in the US the scholar was shocked to find an American mass culture that perpetuated capitalist ideals. Adorno set about critiquing mass culture in terms of the monolithic nature of its ideological and aesthetic continuities and its capitalist and industrial agenda.

Daniel Bell (1919–2011) was an American sociologist whose work concentrated on post-industrialism (the legacies of the period in which industry was a defining influence) in American culture and society. His key works include *The End of Ideology* (1960), *The Coming of Post-Industrialist Society* (1973) and *The Cultural Contradictions of Capitalism* (1976).

Ruth Benedict (1887–1948) was an American cultural anthropologist whose work was interested in cultural groups and cultural organization. Her key works include *The Chrysanthemum and the Sword: Patterns of Japanese Culture* (1946) and *Patterns of Culture* (1934).

Norman Birnbaum (b. 1926) is an American sociologist and professor emeritus at Georgetown University Law Center. His work examines social reform, politics, religion, and all aspects of society.

Louis D. Brandeis (1856–1941) was an American lawyer. President Woodrow Wilson appointed him in 1916 to the Supreme Court, where he served as an associate lawyer until 1939.

Paul Buhle (b. 1944) is a Jewish American scholar whose work focuses on Jewish history, popular culture, radicalism, and American identity.

Holden Caulfield is a fictional character who features in J. D. Salinger's 1951 novel *The Catcher in the Rye*.

Reuel Denney (1913–95) was an American poet and academic who served as professor emeritus at the University of Hawaii.

Bob Dylan (b. 1941) is an American singer-songwriter. He made reference to the "lonely crowd" in his 1967 song "I Shall Be Released."

Erik H. Erikson (1902–94) was a German American psychoanalyst who served as a professor at numerous Ivy League institutions. He is best known for coining the phrase "identity crisis."

Sigmund Freud (1856–1939) was an Austrian neurologist who founded psychoanalysis and wrote extensively on the subject, and on theories of the unconscious, sexuality, repression, and the libido, including: *Three Essays on the Theory of Sexuality* (1905), *The Interpretation of Dreams* (1899) and *Beyond the Pleasure Principle* (1920).

Erich Fromm (1900–80) was a German psychoanalyst and sociologist. Fromm's work looked at the relationship between an individual's social environment and experience and his or her psyche, with a particular interest in consumer culture.

Herbert J. Gans (b. 1927) is an American sociologist, once serving as the president of the American Sociological Association.

Todd Gitlin (b. 1943) is a professor of journalism and sociology at Columbia University. He specializes in the study of activism, mass media, race, nationalism, and politics.

Nathan Glazer (b. 1923) is an American sociologist whose work focuses on ethnicity, race, and cultural inclusion within the United States.

Robert Gutman (1926–2007) was an American sociologist whose work discussed architecture and housing, most famously in *The Design of American Housing* (1985).

Martin Halliwell is a professor of American studies at the University of Leicester. His work examines American culture throughout the twentieth century and especially popular culture, film, and literature since 1945.

Daniel Horowitz is an American historian of consumer culture in the twentieth-century United States. Professor emeritus at Smith College in Northampton, Massachusetts, his books include *The Anxieties of Affluence: Critiques of American Consumer Culture, 1939–1979* (2004).

Christopher Jencks (b. 1936) is an American social scientist who has taught at Northwestern University and the University of Chicago, and is currently the Malcolm Wiener professor of social policy at Harvard University.

Abram Kardiner (1891–1981) was an American anthropologist whose work *The Traumatic Neuroses of War* (1941) was among the first to discuss the psychological trauma of war.

Gideon Lewis-Kraus (b. 1980) is a writer who has written for a number of publications including *Harper's*, the *New York Times Book Review*, the *Los Angeles Times Book Review* and the *New Yorker*. He is the author of *A Sense of Direction* (2014).

Seymour Martin Lipset (1922–2006) was an American sociologist whose work focused on democracy and American exceptionalism (the theory that the United States, with its commitment to personal freedom and democratic ideals, is different to other countries).

Leo Lowenthal (1900–93) was a German sociologist who taught at the University of California, Berkeley.

Sidney Lumet (1924–2011) was a Jewish American film director, scriptwriter, and producer whose most notable work includes *12 Angry Men* (1957), *Dog Day Afternoon* (1975), *Network* (1976), and *The Verdict* (1982).

Robert S. Lynd and Helen Merrell Lynd are a husband-and-wife team who produced the ground-breaking study *Middletown: A Study in Contemporary American Culture* in 1929. Their text addressed key cultural norms and practices in an average American urban enclave.

Wilfred M. McClay is a conservative American historian.

Neil McLaughlin is an academic in the field of sociology.

Herbert Marcuse (1898–1979) was a German philosopher and sociologist whose works include *Reason and Revolution: Hegel and the Rise of Social Theory* (1941) and *One-Dimensional Man* (1964).

Margaret Mead (1901–78) was an American cultural anthropologist who wrote and published widely, looking at areas such as the women's movement, the bomb, primitive societies, and the widening of sexual mores. Her key works include *Coming of Age in Samoa* (1928) and *Sex and Temperament in Three Primitive Societies* (1935).

Vance Packard (1914–96) was an American journalist and social critic who published a series of sociological studies in the late 1950s and 1960s, the best known of which is *The Hidden Persuaders* (1957), which looks at the advertising industry.

William Palmer is a professor of history at Marshall University in Huntington, West Virginia.

Talcott Parsons (1902–79) was an American social theorist whose work included *The Structure of Social Action* (1937) and *The Social System* (1951).

Orlando Patterson (b. 1940) is a Jamaican historian whose work focuses on the issue of race in the United States.

Philip Rieff (1922–2006) was an American sociologist and cultural critic whose work is first and foremost concerned with the analysis of Sigmund Freud.

J. D. Salinger (1919–2010) was an American fiction writer whose best-known works are *The Catcher in the Rye* (1951) and *Franny and Zooey* (1961). He spent much of his later life as a recluse and shunned the limelight that his literary success invited.

Robert T. Sandin is a scholar whose works focuses on religion and conflict.

Lee Siegel (b. 1957) is an American cultural critic whose work has appeared in the *New York Review of Books* and the *New Yorker*. Among his books are *Not Remotely Controlled: Notes on Television* (2007) and *Harvard Is Burning* (2011).

Alexis de Tocqueville (1805–59) was a French political thinker and historian who travelled extensively in the United States. He published his experiences and insights in the seminal study *Democracy in America* (1835, 1840), a book considered to be an authoritative study of the United States in the nineteenth century and one of the first books of sociology.

Thorstein Veblen (1857–1929) was an American sociologist and economist. His most famous work, *The Theory of the Leisure Class* (1899), examined the life of the elite and wealthy members of society and won praise for its engaging and literary style.

William H. Whyte (1917–99) was an American urbanist, journalist, and writer. His 1956 bestseller *The Organization Man* examined American corporate culture and work life in the mid-twentieth century. Whyte also had an interest in urban spaces and published popular research on the subject based on his careful observations.

Sloan Wilson (1920–2003) was an American reporter and novelist who wrote 15 books. Other than *The Man in the Grey Flannel Suit* (1955), which is by far and away Wilson's most successful novel, the author also wrote *A Summer Place* (1958).

Alan Wolfe (b. 1942) is professor of political science and director of the Boisi Center for Religion and American Public Life at Boston College. He specializes in American political and national identity.

C. Wright Mills (1916–62) was an American sociologist, social commentator, and cultural critic, all of which converged to make him a radical social theorist. His most famous work, *The Power Elite* (1956), described the tightly woven and hermetic nature of the ruling classes among the military, in politics, economics and in the top rungs of American socioeconomic life more generally.

Dennis H. Wrong (b. 1923) is an American sociologist best known for his writing on political and social power.

WORKS CITED

WORKS CITED

Birnbaum, Norman. *Toward a Critical Sociology*. Oxford: Oxford University Press, 1973.

Buhle, Paul. "Obituaries: David Riesman." The *Guardian*, May 13, 2002. Accessed June 15, 2013. www.guardian.co.uk/news/2002/may/13/ guardianobituaries.socialsciences .

Epstein, Edwin M. "Religion and Business—The Critical Role of Religious Traditions in Management Education." *Journal of Business Ethics* vol. 38, nos. 1–2 (2002): 91–96.

Featherstone, Joseph. "John Dewey and David Riesman: From the Lost Individual to the Lonely Crowd." *John Dewey: Political Theory and Social Practice* 2 (1992): 59.

Fromm, Erich. *Escape from Freedom*. New York, Farrar & Rinehart, 1941.

Man for Himself: An Inquiry into the Psychology of Ethics. London: Routledge, 2013.

Gans, Herbert J. "Best-sellers by American Sociologists: An Exploratory Study." In Required Reading: Sociology's Most Influential Books, edited by Dan Clawson, 19–27. Amherst: University of Massachusetts Press, 1998.

Gitlin, Todd. Foreword to *The Lonely Crowd: A Study of the Changing American Character* by David Riesman, Nathan Glazer, and Reuel Denney. New Haven: Yale University Press, 2001.

Halliwell, Martin. *American Culture in the 1950s.* Edinburgh: Edinburgh University Press, 2007.

Horowitz, Daniel. "David Riesman: From Law to Social Criticism." *Buffalo Law Review* 58 (2010): 1005.

Jancovich, Mark. *Rational Fears: American Horror in the 1950s*. Manchester: Manchester University Press, 1996.

Jencks, Christopher and David Riesman. The Academic Revolution. Garden City, NY: Doubleday, 1968.

Kassarjian, Waltraud Marggraff. 1962. "A Study of Riesman's Theory of Social Character." *Sociometry* vol. 25 no. 3 (1962): 213–30. Accessed December 15, 2015. doi:10.2307/2786125.

Lewis-Kraus, Gideon. "Yelp and the Wisdom of *The Lonely Crowd*." The *New Yorker.* May 7, 2013. Accessed June 15, 2013. www.newyorker.com/online/ blogs/elements/2013/05/the-wisdom-of-the-lonely-crowds.html.

Lipset, Seymour Martin and Leo Lowenthal, eds. *Culture and Social Character: The Work of David Riesman Reviewed* (New York: Free Press, 1961).

McClay, Wilfred M. "Fifty Years of *The Lonely Crowd*." *The Wilson Quarterly* vol. 22 no. 3 (Summer 1998): 34–42. Accessed December 15, 2015. http://archive.wilsonquarterly.com/sites/default/files/articles/WQ_VOL22_SU_1998_Article_02.pdf.

McGrath, Charles. "The Lives They Lived; Big Thinkster." The *New York Times*, December 29, 2002. Accessed June 15, 2013. www.nytimes.com/2002/12/29/magazine/the-lives-they-lived-big-thinkster.html.

McLaughlin, Neil. "Critical Theory Meets America: Riesman, Fromm, and *The Lonely Crowd*." *The American Sociologist* vol. 32, no. 1 (2001): 5–26.

Nicholson, Ian. "'Shocking' Masculinity: Stanley Milgram, 'Obedience to Authority,' and the 'Crisis of Manhood' in Cold War America." *Isis* vol. 102, no. 2 (2011): 238–68.

Packard, Vance. *The Status Seekers: An Exploration of Class Behavior in America*. Harmondsworth: Penguin, 1961.

Palmer, William. "David Riesman, Alexis de Tocqueville and History: A Look at *The Lonely Crowd* after Forty Years." *Colby Quarterly* vol. 26, no. 1 (1990): 19–27.

Patterson, Orlando. "The Last Sociologist." The *New York Times*, May 19, 2002. Accessed June 15, 2013. www.nytimes.com/2002/05/19/opinion/the-last-sociologist.html.

Riesman, David. *Abundance for What? and Other Essays*. Garden City, NY: Doubleday, 1964.

Individualism Reconsidered and Other Essays. Glencoe, Illinois: Free Press, 1954.

Riesman, David and Nathan Glazer. *Faces in the Crowd: Individual Studies in Character and Politics*. New Haven: Yale University Press, 1952.

"The Meaning of Opinion." *The Public Opinion Quarterly* vol. 12, no. 4 (1948–49): 633–48.

Riesman, David, Nathan Glazer, and Reuel Denney. *The Lonely Crowd: A Study of the Changing American Character*. New Haven: Yale University Press, 2001.

Sandin, Robert T. "Reflections on David Riesman (1909–2002)." *The Pietisten* vol. XVII, no. 1 (2002). Accessed June 15, 2013. www.pietisten.org/summer02/riesman.html.

Siegel, Lee. "Go the Same Way, or Go the Wrong Way." The *New York Times*, May 3, 2013. Accessed June 15, 2013. www.nytimes.com/2013/05/05/

fashion/seeking-out-peer-pressure.html?pagewanted=all&_r=0.

Whyte, William Hollingsworth. The Organization Man. New York: Simon & Schuster, 1956.

Wolfe, Alan. *Marginalized in the Middle.* Chicago: University of Chicago Press, 1996.

Wright Mills, C.. *The Power Elite*. New York: Oxford University Press, 1956.

White Collar: The American Middle Classes. New York: Oxford University Press, 1951.

THE MACAT LIBRARY
BY DISCIPLINE

AFRICANA STUDIES

Chinua Achebe's *An Image of Africa: Racism in Conrad's Heart of Darkness*
W. E. B. Du Bois's *The Souls of Black Folk*
Zora Neale Huston's *Characteristics of Negro Expression*
Martin Luther King Jr's *Why We Can't Wait*
Toni Morrison's *Playing in the Dark: Whiteness in the American Literary Imagination*

ANTHROPOLOGY

Arjun Appadurai's *Modernity at Large: Cultural Dimensions of Globalisation*
Philippe Ariès's *Centuries of Childhood*
Franz Boas's *Race, Language and Culture*
Kim Chan & Renée Mauborgne's *Blue Ocean Strategy*
Jared Diamond's *Guns, Germs & Steel: the Fate of Human Societies*
Jared Diamond's *Collapse: How Societies Choose to Fail or Survive*
E. E. Evans-Pritchard's *Witchcraft, Oracles and Magic Among the Azande*
James Ferguson's *The Anti-Politics Machine*
Clifford Geertz's *The Interpretation of Cultures*
David Graeber's *Debt: the First 5000 Years*
Karen Ho's *Liquidated: An Ethnography of Wall Street*
Geert Hofstede's *Culture's Consequences: Comparing Values, Behaviors, Institutes and Organizations across Nations*
Claude Lévi-Strauss's *Structural Anthropology*
Jay Macleod's *Ain't No Makin' It: Aspirations and Attainment in a Low-Income Neighborhood*
Saba Mahmood's *The Politics of Piety: The Islamic Revival and the Feminist Subjec*t
Marcel Mauss's *The Gift*

BUSINESS

Jean Lave & Etienne Wenger's *Situated Learning*
Theodore Levitt's *Marketing Myopia*
Burton G. Malkiel's *A Random Walk Down Wall Street*
Douglas McGregor's *The Human Side of Enterprise*
Michael Porter's *Competitive Strategy: Creating and Sustaining Superior Performance*
John Kotter's *Leading Change*
C. K. Prahalad & Gary Hamel's *The Core Competence of the Corporation*

CRIMINOLOGY

Michelle Alexander's *The New Jim Crow: Mass Incarceration in the Age of Colorblindness*
Michael R. Gottfredson & Travis Hirschi's *A General Theory of Crime*
Richard Herrnstein & Charles A. Murray's *The Bell Curve: Intelligence and Class Structure in American Life*
Elizabeth Loftus's *Eyewitness Testimony*
Jay Macleod's *Ain't No Makin' It: Aspirations and Attainment in a Low-Income Neighborhood*
Philip Zimbardo's *The Lucifer Effect*

ECONOMICS

Janet Abu-Lughod's *Before European Hegemony*
Ha-Joon Chang's *Kicking Away the Ladder*
David Brion Davis's *The Problem of Slavery in the Age of Revolution*
Milton Friedman's *The Role of Monetary Policy*
Milton Friedman's *Capitalism and Freedom*
David Graeber's *Debt: the First 5000 Years*
Friedrich Hayek's *The Road to Serfdom*
Karen Ho's *Liquidated: An Ethnography of Wall Street*

John Maynard Keynes's *The General Theory of Employment, Interest and Money*
Charles P. Kindleberger's *Manias, Panics and Crashes*
Robert Lucas's *Why Doesn't Capital Flow from Rich to Poor Countries?*
Burton G. Malkiel's *A Random Walk Down Wall Street*
Thomas Robert Malthus's *An Essay on the Principle of Population*
Karl Marx's *Capital*
Thomas Piketty's *Capital in the Twenty-First Century*
Amartya Sen's *Development as Freedom*
Adam Smith's *The Wealth of Nations*
Nassim Nicholas Taleb's *The Black Swan: The Impact of the Highly Improbable*
Amos Tversky's & Daniel Kahneman's *Judgment under Uncertainty: Heuristics and Biases*
Mahbub Ul Haq's *Reflections on Human Development*
Max Weber's *The Protestant Ethic and the Spirit of Capitalism*

FEMINISM AND GENDER STUDIES

Judith Butler's *Gender Trouble*
Simone De Beauvoir's *The Second Sex*
Michel Foucault's *History of Sexuality*
Betty Friedan's *The Feminine Mystique*
Saba Mahmood's *The Politics of Piety: The Islamic Revival and the Feminist Subject*
Joan Wallach Scott's *Gender and the Politics of History*
Mary Wollstonecraft's *A Vindication of the Rights of Woman*
Virginia Woolf's *A Room of One's Own*

GEOGRAPHY

The Brundtland Report's *Our Common Future*
Rachel Carson's *Silent Spring*
Charles Darwin's *On the Origin of Species*
James Ferguson's *The Anti-Politics Machine*
Jane Jacobs's *The Death and Life of Great American Cities*
James Lovelock's *Gaia: A New Look at Life on Earth*
Amartya Sen's *Development as Freedom*
Mathis Wackernagel & William Rees's *Our Ecological Footprint*

HISTORY

Janet Abu-Lughod's *Before European Hegemony*
Benedict Anderson's *Imagined Communities*
Bernard Bailyn's *The Ideological Origins of the American Revolution*
Hanna Batatu's *The Old Social Classes And The Revolutionary Movements Of Iraq*
Christopher Browning's *Ordinary Men: Reserve Police Batallion 101 and the Final Solution in Poland*
Edmund Burke's *Reflections on the Revolution in France*
William Cronon's *Nature's Metropolis: Chicago And The Great West*
Alfred W. Crosby's *The Columbian Exchange*
Hamid Dabashi's *Iran: A People Interrupted*
David Brion Davis's *The Problem of Slavery in the Age of Revolution*
Nathalie Zemon Davis's *The Return of Martin Guerre*
Jared Diamond's *Guns, Germs & Steel: the Fate of Human Societies*
Frank Dikotter's *Mao's Great Famine*
John W Dower's *War Without Mercy: Race And Power In The Pacific War*
W. E. B. Du Bois's *The Souls of Black Folk*
Richard J. Evans's *In Defence of History*
Lucien Febvre's *The Problem of Unbelief in the 16th Century*
Sheila Fitzpatrick's *Everyday Stalinism*

Eric Foner's *Reconstruction: America's Unfinished Revolution, 1863-1877*
Michel Foucault's *Discipline and Punish*
Michel Foucault's *History of Sexuality*
Francis Fukuyama's *The End of History and the Last Man*
John Lewis Gaddis's *We Now Know: Rethinking Cold War History*
Ernest Gellner's *Nations and Nationalism*
Eugene Genovese's *Roll, Jordan, Roll: The World the Slaves Made*
Carlo Ginzburg's *The Night Battles*
Daniel Goldhagen's *Hitler's Willing Executioners*
Jack Goldstone's *Revolution and Rebellion in the Early Modern World*
Antonio Gramsci's *The Prison Notebooks*
Alexander Hamilton, John Jay & James Madison's *The Federalist Papers*
Christopher Hill's *The World Turned Upside Down*
Carole Hillenbrand's *The Crusades: Islamic Perspectives*
Thomas Hobbes's *Leviathan*
Eric Hobsbawm's *The Age Of Revolution*
John A. Hobson's *Imperialism: A Study*
Albert Hourani's *History of the Arab Peoples*
Samuel P. Huntington's *The Clash of Civilizations and the Remaking of World Order*
C. L. R. James's *The Black Jacobins*
Tony Judt's *Postwar: A History of Europe Since 1945*
Ernst Kantorowicz's *The King's Two Bodies: A Study in Medieval Political Theology*
Paul Kennedy's *The Rise and Fall of the Great Powers*
Ian Kershaw's *The "Hitler Myth": Image and Reality in the Third Reich*
John Maynard Keynes's *The General Theory of Employment, Interest and Money*
Charles P. Kindleberger's *Manias, Panics and Crashes*
Martin Luther King Jr's *Why We Can't Wait*
Henry Kissinger's *World Order: Reflections on the Character of Nations and the Course of History*
Thomas Kuhn's *The Structure of Scientific Revolutions*
Georges Lefebvre's *The Coming of the French Revolution*
John Locke's *Two Treatises of Government*
Niccolò Machiavelli's *The Prince*
Thomas Robert Malthus's *An Essay on the Principle of Population*
Mahmood Mamdani's *Citizen and Subject: Contemporary Africa And The Legacy Of Late Colonialism*
Karl Marx's *Capital*
Stanley Milgram's *Obedience to Authority*
John Stuart Mill's *On Liberty*
Thomas Paine's *Common Sense*
Thomas Paine's *Rights of Man*
Geoffrey Parker's *Global Crisis: War, Climate Change and Catastrophe in the Seventeenth Century*
Jonathan Riley-Smith's *The First Crusade and the Idea of Crusading*
Jean-Jacques Rousseau's *The Social Contract*
Joan Wallach Scott's *Gender and the Politics of History*
Theda Skocpol's *States and Social Revolutions*
Adam Smith's *The Wealth of Nations*
Timothy Snyder's *Bloodlands: Europe Between Hitler and Stalin*
Sun Tzu's *The Art of War*
Keith Thomas's *Religion and the Decline of Magic*
Thucydides's *The History of the Peloponnesian War*
Frederick Jackson Turner's *The Significance of the Frontier in American History*
Odd Arne Westad's *The Global Cold War: Third World Interventions And The Making Of Our Times*

LITERATURE

Chinua Achebe's *An Image of Africa: Racism in Conrad's Heart of Darkness*
Roland Barthes's *Mythologies*
Homi K. Bhabha's *The Location of Culture*
Judith Butler's *Gender Trouble*
Simone De Beauvoir's *The Second Sex*
Ferdinand De Saussure's *Course in General Linguistics*
T. S. Eliot's *The Sacred Wood: Essays on Poetry and Criticism*
Zora Neale Huston's *Characteristics of Negro Expression*
Toni Morrison's *Playing in the Dark: Whiteness in the American Literary Imagination*
Edward Said's *Orientalism*
Gayatri Chakravorty Spivak's *Can the Subaltern Speak?*
Mary Wollstonecraft's *A Vindication of the Rights of Women*
Virginia Woolf's *A Room of One's Own*

PHILOSOPHY

Elizabeth Anscombe's *Modern Moral Philosophy*
Hannah Arendt's *The Human Condition*
Aristotle's *Metaphysics*
Aristotle's *Nicomachean Ethics*
Edmund Gettier's *Is Justified True Belief Knowledge?*
Georg Wilhelm Friedrich Hegel's *Phenomenology of Spirit*
David Hume's *Dialogues Concerning Natural Religion*
David Hume's *The Enquiry for Human Understanding*
Immanuel Kant's *Religion within the Boundaries of Mere Reason*
Immanuel Kant's *Critique of Pure Reason*
Søren Kierkegaard's *The Sickness Unto Death*
Søren Kierkegaard's *Fear and Trembling*
C. S. Lewis's *The Abolition of Man*
Alasdair MacIntyre's *After Virtue*
Marcus Aurelius's *Meditations*
Friedrich Nietzsche's *On the Genealogy of Morality*
Friedrich Nietzsche's *Beyond Good and Evil*
Plato's *Republic*
Plato's *Symposium*
Jean-Jacques Rousseau's *The Social Contract*
Gilbert Ryle's *The Concept of Mind*
Baruch Spinoza's *Ethics*
Sun Tzu's *The Art of War*
Ludwig Wittgenstein's *Philosophical Investigations*

POLITICS

Benedict Anderson's *Imagined Communities*
Aristotle's *Politics*
Bernard Bailyn's *The Ideological Origins of the American Revolution*
Edmund Burke's *Reflections on the Revolution in France*
John C. Calhoun's *A Disquisition on Government*
Ha-Joon Chang's *Kicking Away the Ladder*
Hamid Dabashi's *Iran: A People Interrupted*
Hamid Dabashi's *Theology of Discontent: The Ideological Foundation of the Islamic Revolution in Iran*
Robert Dahl's *Democracy and its Critics*
Robert Dahl's *Who Governs?*
David Brion Davis's *The Problem of Slavery in the Age of Revolution*

Alexis De Tocqueville's *Democracy in America*
James Ferguson's *The Anti-Politics Machine*
Frank Dikotter's *Mao's Great Famine*
Sheila Fitzpatrick's *Everyday Stalinism*
Eric Foner's *Reconstruction: America's Unfinished Revolution, 1863-1877*
Milton Friedman's *Capitalism and Freedom*
Francis Fukuyama's *The End of History and the Last Man*
John Lewis Gaddis's *We Now Know: Rethinking Cold War History*
Ernest Gellner's *Nations and Nationalism*
David Graeber's *Debt: the First 5000 Years*
Antonio Gramsci's *The Prison Notebooks*
Alexander Hamilton, John Jay & James Madison's *The Federalist Papers*
Friedrich Hayek's *The Road to Serfdom*
Christopher Hill's *The World Turned Upside Down*
Thomas Hobbes's *Leviathan*
John A. Hobson's *Imperialism: A Study*
Samuel P. Huntington's *The Clash of Civilizations and the Remaking of World Order*
Tony Judt's *Postwar: A History of Europe Since 1945*
David C. Kang's *China Rising: Peace, Power and Order in East Asia*
Paul Kennedy's *The Rise and Fall of Great Powers*
Robert Keohane's *After Hegemony*
Martin Luther King Jr.'s *Why We Can't Wait*
Henry Kissinger's *World Order: Reflections on the Character of Nations and the Course of History*
John Locke's *Two Treatises of Government*
Niccolò Machiavelli's *The Prince*
Thomas Robert Malthus's *An Essay on the Principle of Population*
Mahmood Mamdani's *Citizen and Subject: Contemporary Africa And The Legacy Of Late Colonialism*
Karl Marx's *Capital*
John Stuart Mill's *On Liberty*
John Stuart Mill's *Utilitarianism*
Hans Morgenthau's *Politics Among Nations*
Thomas Paine's *Common Sense*
Thomas Paine's *Rights of Man*
Thomas Piketty's *Capital in the Twenty-First Century*
Robert D. Putnam's *Bowling Alone*
John Rawls's *Theory of Justice*
Jean-Jacques Rousseau's *The Social Contract*
Theda Skocpol's *States and Social Revolutions*
Adam Smith's *The Wealth of Nations*
Sun Tzu's *The Art of War*
Henry David Thoreau's *Civil Disobedience*
Thucydides's *The History of the Peloponnesian War*
Kenneth Waltz's *Theory of International Politics*
Max Weber's *Politics as a Vocation*
Odd Arne Westad's *The Global Cold War: Third World Interventions And The Making Of Our Times*

POSTCOLONIAL STUDIES

Roland Barthes's *Mythologies*
Frantz Fanon's *Black Skin, White Masks*
Homi K. Bhabha's *The Location of Culture*
Gustavo Gutiérrez's *A Theology of Liberation*
Edward Said's *Orientalism*
Gayatri Chakravorty Spivak's *Can the Subaltern Speak?*

PSYCHOLOGY

Gordon Allport's *The Nature of Prejudice*
Alan Baddeley & Graham Hitch's *Aggression: A Social Learning Analysis*
Albert Bandura's *Aggression: A Social Learning Analysis*
Leon Festinger's *A Theory of Cognitive Dissonance*
Sigmund Freud's *The Interpretation of Dreams*
Betty Friedan's *The Feminine Mystique*
Michael R. Gottfredson & Travis Hirschi's *A General Theory of Crime*
Eric Hoffer's *The True Believer: Thoughts on the Nature of Mass Movements*
William James's *Principles of Psychology*
Elizabeth Loftus's *Eyewitness Testimony*
A. H. Maslow's *A Theory of Human Motivation*
Stanley Milgram's *Obedience to Authority*
Steven Pinker's *The Better Angels of Our Nature*
Oliver Sacks's *The Man Who Mistook His Wife For a Hat*
Richard Thaler & Cass Sunstein's *Nudge: Improving Decisions About Health, Wealth and Happiness*
Amos Tversky's *Judgment under Uncertainty: Heuristics and Biases*
Philip Zimbardo's *The Lucifer Effect*

SCIENCE

Rachel Carson's *Silent Spring*
William Cronon's *Nature's Metropolis: Chicago And The Great West*
Alfred W. Crosby's *The Columbian Exchange*
Charles Darwin's *On the Origin of Species*
Richard Dawkin's *The Selfish Gene*
Thomas Kuhn's *The Structure of Scientific Revolutions*
Geoffrey Parker's *Global Crisis: War, Climate Change and Catastrophe in the Seventeenth Century*
Mathis Wackernagel & William Rees's *Our Ecological Footprint*

SOCIOLOGY

Michelle Alexander's *The New Jim Crow: Mass Incarceration in the Age of Colorblindness*
Gordon Allport's *The Nature of Prejudice*
Albert Bandura's *Aggression: A Social Learning Analysis*
Hanna Batatu's *The Old Social Classes And The Revolutionary Movements Of Iraq*
Ha-Joon Chang's *Kicking Away the Ladder*
W. E. B. Du Bois's *The Souls of Black Folk*
Émile Durkheim's *On Suicide*
Frantz Fanon's *Black Skin, White Masks*
Frantz Fanon's *The Wretched of the Earth*
Eric Foner's *Reconstruction: America's Unfinished Revolution, 1863-1877*
Eugene Genovese's *Roll, Jordan, Roll: The World the Slaves Made*
Jack Goldstone's *Revolution and Rebellion in the Early Modern World*
Antonio Gramsci's *The Prison Notebooks*
Richard Herrnstein & Charles A Murray's *The Bell Curve: Intelligence and Class Structure in American Life*
Eric Hoffer's *The True Believer: Thoughts on the Nature of Mass Movements*
Jane Jacobs's *The Death and Life of Great American Cities*
Robert Lucas's *Why Doesn't Capital Flow from Rich to Poor Countries?*
Jay Macleod's *Ain't No Makin' It: Aspirations and Attainment in a Low Income Neighborhood*
Elaine May's *Homeward Bound: American Families in the Cold War Era*
Douglas McGregor's *The Human Side of Enterprise*
C. Wright Mills's *The Sociological Imagination*

Thomas Piketty's *Capital in the Twenty-First Century*
Robert D. Putman's *Bowling Alone*
David Riesman's *The Lonely Crowd: A Study of the Changing American Character*
Edward Said's *Orientalism*
Joan Wallach Scott's *Gender and the Politics of History*
Theda Skocpol's *States and Social Revolutions*
Max Weber's *The Protestant Ethic and the Spirit of Capitalism*

THEOLOGY

Augustine's *Confessions*
Benedict's *Rule of St Benedict*
Gustavo Gutiérrez's *A Theology of Liberation*
Carole Hillenbrand's *The Crusades: Islamic Perspectives*
David Hume's *Dialogues Concerning Natural Religion*
Immanuel Kant's *Religion within the Boundaries of Mere Reason*
Ernst Kantorowicz's *The King's Two Bodies: A Study in Medieval Political Theology*
Søren Kierkegaard's *The Sickness Unto Death*
C. S. Lewis's *The Abolition of Man*
Saba Mahmood's *The Politics of Piety: The Islamic Revival and the Feminist Subject*
Baruch Spinoza's *Ethics*
Keith Thomas's *Religion and the Decline of Magic*

COMING SOON

Chris Argyris's *The Individual and the Organisation*
Seyla Benhabib's *The Rights of Others*
Walter Benjamin's *The Work Of Art in the Age of Mechanical Reproduction*
John Berger's *Ways of Seeing*
Pierre Bourdieu's *Outline of a Theory of Practice*
Mary Douglas's *Purity and Danger*
Roland Dworkin's *Taking Rights Seriously*
James G. March's *Exploration and Exploitation in Organisational Learning*
Ikujiro Nonaka's *A Dynamic Theory of Organizational Knowledge Creation*
Griselda Pollock's *Vision and Difference*
Amartya Sen's *Inequality Re-Examined*
Susan Sontag's *On Photography*
Yasser Tabbaa's *The Transformation of Islamic Art*
Ludwig von Mises's *Theory of Money and Credit*

Macat Pairs

Analyse historical and modern issues from opposite sides of an argument. Pairs include:

RACE AND IDENTITY

Zora Neale Hurston's
Characteristics of Negro Expression

Using material collected on anthropological expeditions to the South, Zora Neale Hurston explains how expression in African American culture in the early twentieth century departs from the art of white America. At the time, African American art was often criticized for copying white culture. For Hurston, this criticism misunderstood how art works. European tradition views art as something fixed. But Hurston describes a creative process that is alive, ever-changing, and largely improvisational. She maintains that African American art works through a process called 'mimicry'—where an imitated object or verbal pattern, for example, is reshaped and altered until it becomes something new, novel—and worthy of attention.

Frantz Fanon's
Black Skin, White Masks

Black Skin, White Masks offers a radical analysis of the psychological effects of colonization on the colonized.

Fanon witnessed the effects of colonization first hand both in his birthplace, Martinique, and again later in life when he worked as a psychiatrist in another French colony, Algeria. His text is uncompromising in form and argument. He dissects the dehumanizing effects of colonialism, arguing that it destroys the native sense of identity, forcing people to adapt to an alien set of values—including a core belief that they are inferior. This results in deep psychological trauma.

Fanon's work played a pivotal role in the civil rights movements of the 1960s.

Macat analyses are available from all good bookshops and libraries.

Access hundreds of analyses through one, multimedia tool.
Join free for one month **library.macat.com**

Macat Pairs

*Analyse historical and modern issues
from opposite sides of an argument.
Pairs include:*

MACAT

MACAT

INTERNATIONAL RELATIONS IN THE 21ˢᵀ CENTURY

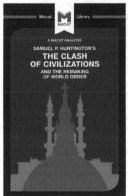

Samuel P. Huntington's
The Clash of Civilisations

In his highly influential 1996 book, Huntington offers a vision of a post-Cold War world in which conflict takes place not between competing ideologies but between cultures. The worst clash, he argues, will be between the Islamic world and the West: the West's arrogance and belief that its culture is a "gift" to the world will come into conflict with Islam's obstinacy and concern that its culture is under attack from a morally decadent "other."

Clash inspired much debate between different political schools of thought. But its greatest impact came in helping define American foreign policy in the wake of the 2001 terrorist attacks in New York and Washington.

Francis Fukuyama's
The End of History and the Last Man

Published in 1992, *The End of History and the Last Man* argues that capitalist democracy is the final destination for all societies. Fukuyama believed democracy triumphed during the Cold War because it lacks the "fundamental contradictions" inherent in communism and satisfies our yearning for freedom and equality. Democracy therefore marks the endpoint in the evolution of ideology, and so the "end of history." There will still be "events," but no fundamental change in ideology.

Macat Pairs

Analyse historical and modern issues from opposite sides of an argument. Pairs include:

Printed in the United States
by Baker & Taylor Publisher Services